Protoplasmatologia

Begründet von / Founded by

L. V. Heilbrunn, Philadelphia, Pa. · F. Weber, Graz

Herausgegeben von / Edited by

M. Alfert, Berkeley, Calif.
W. Beermann, Tübingen

W. Sandritter, Freiburg i. Br.
P. Sitte, Freiburg i. Br.

Mitherausgeber / Advisory Board

J. Brachet, Bruxelles
D. Branton, Berkeley, Calif.
H. G. Callan, St. Andrews
W. W. Franke, Freiburg i. Br.

N. Kamiya, Osaka
G. F. Springer, Evanston, Ill.
L. Stockinger, Wien
B. F. Trump, Baltimore, Md.

III Cytoplasmic Organelles
G Comparative Organellography
   of the Cytoplasm

Springer-Verlag
Wien New York 1973

Comparative Organellography of the Cytoplasm

A. Frey-Wyssling

With 31 Figures and 9 Plates

Springer-Verlag
Wien New York 1973

© 1973 by Springer-Verlag/Wien
Softcover reprint of the hardcover 1st edition 1973
Library of Congress Catalog Card Number 73-75261

ISBN-13: 978-3-7091-5614-8   e-ISBN-13: 978-3-7091-5612-4
DOI: 10.1007/978-3-7091-5612-4

Vor 20 Jahren haben Professor FRIEDL WEBER, Graz, und Professor L. V. HEILBRUNN, Philadelphia, das Handbuch „Protoplasmatologia" gegründet, zu einer Zeit, als es der Stand der Wissenschaft noch erlaubte, an ein Handbuch im klassischen Sinne zu denken, das heißt an eine umfassende Darstellung des Gesamtgebietes. Die ursprüngliche Disposition sah zwölf Bände mit einer hierarchischen Untergliederung vor. Die rasche Entwicklung auf diesem Gebiet mit der zunehmenden Differenzierung einerseits und der Bildung neuer Schwerpunkte andererseits hat im Lauf der Zeit mehrfache Änderungen der Disposition notwendig gemacht. Neue Gesichtspunkte ergaben sich auch durch den Wechsel im Herausgeberkollegium nach dem Tod der Gründer.

Seit dem Jahre 1953 sind 52 Einzelbände mit einem Gesamtumfang von rund 9400 Seiten erschienen.

Die Herausgeber haben nun im Einvernehmen mit dem Verlag beschlossen, die engen Schranken der früheren Handbuchdisposition zu verlassen. In einer zeitgemäßeren Form sollen Ergebnisse und Probleme der Zellbiologie in Monographien dargestellt werden. So wird es möglich sein, jeweils besonders aktuelle Themen zu behandeln. Erhalten bleiben wird der Anspruch auf höchstes wissenschaftliches Niveau.

Herausgeber und Verlag

Twenty years ago, Prof. FRIEDL WEBER (Graz University) and Prof. L. V. HEILBRUNN (University of Pennsylvania) conceived the idea for the handbook "Protoplasmatologia" at a time when the state of knowledge in the field of cell biology still permitted one to think of an all-encompassing handbook in the classical sense. Since 1953 fifty-two volumes with a total of about 9,400 pages have been published. The very rapid developments in this area of science, especially during the last decade, have led to new insights which necessitated some alterations in the original plan of the handbook; also, changes in the board of editors since the death of the founders have brought about a reorientation of viewpoints.

The editors, in agreement with the publisher, have now decided to abandon the confining limits of the original disposition of the handbook altogether and to continue this work, in a form more appropriate to current needs, as an open series of monographs dealing with present-day problems and findings in cell biology. This will make it possible to treat the most modern and interesting aspects of the field as they arise in the course of contemporary research. The highest scientific, editorial and publishing standards will continue to be maintained.

Editors and publisher

Preface

Since the publication of my monograph "Die submikroskopische Struktur des Cytoplasmas" [Protoplasmatologia II/A/2 (1955)], science has increased our knowledge on this cell constituent in an amazing way. At that time, cytoplasm was still considered as an amorphous colloid with an amicroscopic structure capable of sol/gel transformations. Meanwhile, within the past 18 years, ultrastructure research and molecular biology have shown that it consists of structured organelles, thereby revolutionizing our concept of the cytoplasm.

There is an ontogeny of these organelles which deals with their appearance, development, growth and eventual disappearance. Although the results of relevant studies are less spectacular at this time than those concerning nucleic acids, they will become more and more important in the near future, because they touch the basic question of how molecular morphogenesis is realized. Therefore, an attempt is made to portray the actual state of these problems from a biological point of view which involves a comparison of the established organelles.

I thank Miss SONIA TURLER and Dr. ELSA HÄUSERMANN for their help with the manuscript and my former pupils and assistants Drs. KOPP, KUHN, SCHWARZENBACH, STAEHELIN, STEINMANN, and VOGEL, as well as my colleagues Professors MATILE, MOOR, and MÜHLETHALER for original electron micrographs.

Zürich (Switzerland), April 1973   A. FREY-WYSSLING

Comparative Organellography of the Cytoplasm

By

A. Frey-Wyssling

Professor emeritus at the Department of General Botany, Swiss Federal Institute
of Technology Zürich (ETHZ)

With 31 Figures and 9 Plates

Contents

Abbreviations

ATP	Adenosine triphosphate
Chl	Chloroplast
CW	Cell wall
Cy	Cytoplasm
DIP	Dichlorophenol indophenol
DNA	Deoxyribose nucleic acid
DNas	Deoxyribo-nuclease
ER	Endoplasmic reticulum
GA	Golgi apparatus
GP	Groundplasm
gr	Granule
GTP	Guanosine triphosphate
GV	Golgi vesicles
Mi	Mitochondrion
m-RNA	Messenger RNA
MT (Mt)	Microtubule
NAD	Nicotinamide adenine dinucleotide
NADH	Reduced NAD
NADP	Nicotinamide adenine dinucleotide phosphate
NADPH	Reduced NADP
NE	Nuclear envelope
Nu	Nucleus, nuclear
PAS	Periodic acid Schiff reagent
PL	Plasmalemma
Pl	Plastid
rb	Ribosome
RNA	Ribose nucleic acid
RNase	Ribo-nuclease
Sph	Spherosome
TMV	Tobacco Mosaic Virus
t-RNA	Transfer RNA
UDP	Uridine diphosphate
UDPG	UDP glucose
UTP	Uridine triphosphate
Va	Vacuole

Introduction: "Organellography"

The Greek word *"organon"* means a tool. "Organography" is, therefore, the detailed description of the "tools" or organs with which animals and plants (GOEBEL 1913/1923) perform their functions of life. The accounts of organography are based on macroscopic facts and on observations made with the light microscope. With this instrument it was discovered that not only on the histological level but also on the cytological level there are differentiations which guarantee a division of work *inside the cell*.

A short inventory of these differentiations in a plant cell reads as follows:

Protoplast	Nucleus (Nu) Plastids (Pl) Mitochondria (Mi) Cytoplasm (Cy) ———————— Vacuoles (Va) Spherosomes (Sph) Cell wall (CW)	Plasmalemma (PL) Golgi apparatus (GA) Endoplasmic reticulum (ER) Microtubules (MT) Groundplasm (GP)
	light-microscopic dimensions	ultrastructural dimensions

Since specific functions can be attributed to these cell components, they are called *organelles*. The organelles listed on the left have light-microscopic sizes. Together (with the exception of the cell wall which plays a more passive role), they constitute the metabolizing *protoplast*.

It is deplorable that this general term which was originally created to characterize the totality of the living contents in the cell, is misused in recent times for the designation of cells which have been artificially freed of their cell walls. Since the more appropriate term "gymnoplast" (= naked protoplast) exists for such objects, there is no reason for using a well established general term for a special type of mutilated cells (FREY-WYSSLING 1967, STÄHELIN 1954).

Although the *cell wall* is a feature restricted to plant cells, it merits general attention from a historical point of view, because it has given the names to the Cell and to our science Cytology (Greek *kytos* = envelope, coat or even armor). If the prototype of a cell had not been the periderm cell of bottle cork (ROBERT HOOKE 1667) but, for instance, an amoeba, the terms "cell", "cytology" and "cytoplasm" would hardly exist.

The *cytoplasm* deserves special consideration. It is the cell constituent which is left over, when all inclusions visible in the light microscope (nucleus, plastids, mitochondria, spherosomes, vacuoles) are taken away. Thus it is defined by negative criteria, in that it is not said what it really is, but what it does not include as a kind of matrix for the defined cell organelles.

This difficulty of definition became evident when, with the electron microscope, Golgi dictyosomes, the endoplasmic reticulum and microtubules were discovered and cell components of hitherto hypothetical or controversial existence such as the plasmalemma, centrioles and diplosomes became clearly visible.

The *ultrastructural organelles,* listed above on the right-hand side, are also embedded in a matrix and the question arises whether the term "cytoplasm" should be applied to this material in attributing to the Golgi, the reticulum, the microtubules, etc., the same status as to the organelles cited on the left. This is not advisable for the following reasons:

Nuclei, plastids, and mitochondria are separated from the cytoplasm by double membranes called "envelopes" which enclose a periorganellic space while the ultrastructural organelles float freely in the matrix so that it is more difficult to isolate them without plasmic impurities. They also seem to have a lower state of autonomy than plastids and mitochondria.

There is another difficulty: without doubt all the biochemical processes performed by the matrix are bound to special structural systems in a similar way as protein synthesis depends on polysomes (see 3.5.1.). In the same measure as such systems will be detected by molecular biology, the concept "cytoplasm" would become more and more restricted and finally when all functions can be correlated with structural details, would turn out to be superfluous; the "cytoplasm" would just vanish.

For these reasons it seems better to adhere to the classical concept of cytoplasm which comprises all cell constituents with the exception of nuclei, plastids, mitochondria, and cell walls. As a consequence, a new term had to be adopted for the matrix under consideration which harbors the ultrastructural organelles; this term is: *groundplasm.*

The groundplasm has all the structural and functional properties formerly attributed to the cytoplasm (FREY-WYSSLING 1955). It is a colloid which can reversibly adopt the state of a gel or a sol. Depending on the resolution power of the microscope applied, it appears homogeneous or granular. The granules are of the size of macromolecules and their isolation, chemistry, and cooperation offer a vast field of research. The groundplasm may also contain threads of nucleic acid visible in high resolution electron micrographs. Other features are not yet resolvable and still remain amicroscopic.

The functions of the groundplasm are manifold. It is involved in growth phenomena, morphogenesis, plasmic heredity (plasmon), sensorial transmission, etc. It represents a pool in which science locates all kinds of non-clarified functions. Also biochemically it plays an important role and cooperates with the organelles it surrounds. As an example it may be mentioned that the first step of respiration, glycolysis, is performed in the groundplasm, while the subsequent tricarbonic-acid cycle occurs in the mitochondria.

The light-microscopic organelles and the groundplasm cannot be dealt with in this monograph. Of course it would be attractive to include in a Comparative Organellography all cell organelles and especially *mitochondria* and *chloroplasts* which represent landmarks in the history of ultrastructural cytology. They would also furnish instructive examples for speculations on the phylogeny of organelles, e.g. how the plastidal thylakoids (see 1.5.2.) developed in the plastid-free prokaryotes and how such "discs" are wrapped in an envelope in the red algae (Bisalputra 1967, Bouck 1962) as intermediate developmental stages to the chloroplasts of green algae and the higher plants.

However, since the bibliography on plastids and mitochondria is extremely vast and controversial with respect to their autonomy and the size of their self-reproductive units (for plastids: whole chloroplasts, proplastids, initials and/or molecular starters for thylakoids), a presentation of these problems would exceed the scope of this monograph.

The same is true of the ultrastructural problems concerning chromosomes and the nucleus. For these reasons, this monograph is restricted to the organelles of the cytoplasm, including those which outgrow ultrastructural dimensions such as vacuoles and spherosomes.

As a botanist, the author stresses the problems of plant cytology (Frey-Wyssling and Mühlethaler 1965, Clowes and Juniper 1968), and especially those which have hitherto escaped the attention of "general biologists". To my mind "General Cytology" is not only an introductary survey of attractive cytological facts and problems, but in the first place a Comparative Science which should embrace all cytological facts gained by cellular studies on protobionts, animals *and* plants.

Therefore, General Cytology cannot be an introductory course for biologists, however beneficial it is for biochemists and medical and agricultural students, because its cope-stone "Comparative Organellography" can only be studied *after* sufficient experience with both animal and plant cells has been gained.

1. Plasmalemma (PL)

Due to their importance, the *cytomembranes* which guarantee the compartimentation of the cell and separate the protoplast from its surroundings have become the subject of a new science called *membranology*. Special journals such as "Biomembranes" (Amsterdam: Elsevier, four volumes a year) deal with the chemistry, physiology, physics and energetics of these indispensable boundaries within and around the cytoplasm. It is interesting that this fascinating branch of biological research has its origin in investigations on the behavior of the plasma membrane (cell membrane) or *Plasmalemma.*

This boundary film displays the remarkable property of semipermeability, which accounts for plasmolysis, a phenomenon known for more than a century which proves the barrier function of the plasmalemma for molecules larger than those of water. However, the barrier is only maintained as long as the cell is able to supply the membrane with energy. In a dead cell all molecules will diffuse without hindrance through the plasmalemma. This is the reason why only living cells can be plasmolyzed.

Semipermeability is never perfect. Certain molecules may slowly permeate through the plasmalemma and cause deplasmolysis of a plasmolyzed cell. The penetration can be measured and the permeability specified by the so-called coefficient of permeation. Its dimension is cm/sec and differs from the coefficient of diffusion cm²/sec where the length cm appears in the second power. This is due to the fact that for diffusion phenomena the distance Δs

must be known over which the difference of concentration Δc is effective in forming the gradient of diffusion Δc/Δs. For classical cytologists only Δc was measurable, while the width of the plasmalemma as a merely theoretical boundary was completely unknown.

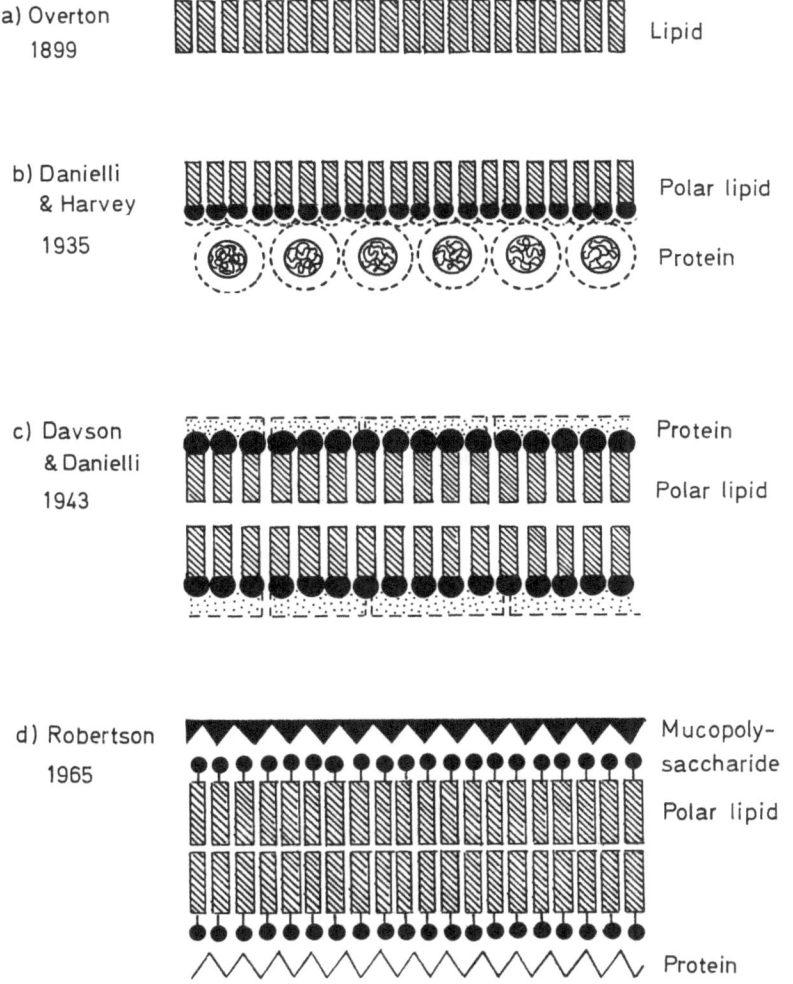

a) Overton 1899 — Lipid

b) Danielli & Harvey 1935 — Polar lipid, Protein

c) Davson & Danielli 1943 — Protein, Polar lipid

d) Robertson 1965 — Mucopolysaccharide, Polar lipid, Protein

Fig. 1. Historical survey of plasmalemma models.

1.1. Ultrastructure

Although the plasmalemma is invisible in the light microscope, conceptions of its structure were developed by indirect methods. OVERTON (1899) who found the value of the coefficient of permeation to rise with the number of lipophilic groups in the penetrating molecules, formulated the so-called lipid-theory. Accordingly the plasmalemma was considered as a lipophilic surface film (Fig. 1 a). However, measurements of the surface tension of a water/oil interface by Danielli and Harvey (1935) yielded a tension ten times higher than that actually found at the interface of fish eggs in their culture medium. Therefore, these authors concluded that the oil phase of the plasmalemma must be covered by a film of globular hydrophilic protein molecules (Fig. 1b). This system would be about 6 nm thick (lipid chains 2.5 nm long + 3.5 nm minimum diameter of globular protein molecules). Yet, such a structure would result in an unstable plasmalemma with a tendency to disintegrate into vesicles, because the surface tension would be too grossly different on either side of such a membrane. This difficulty was overcome by doubling the system and so postulating the wellknown sandwich structure of Fig. 1c (Dawson and Danielli 1943). Although the constant energy consumption of the living plasmalemma suggests an unstable membrane, the static equilibrium model of Fig. 1c was generally accepted.

However, when it became possible to image cross-sections of membranes and to measure their width of 7.5–10 nm in the electron microscope, the chemically fixed plasmalemma proved to be thinner than a sandwich model of 2×6 nm $= 12$ nm. As a consequence it was modified by Robertson (1959). This author stipulated the protein layer as unfolded polypeptic chains (Fig. 1d) or as chains in α-helical conformation. As he claimed this arrangement to be characteristic of all cytomembranes, he declared it a general structure and termed it *unit membrane*.

The concept of the unit membrane is based on the fact that the cytomembranes display three layers in the electron microscope, two of which are electron-dense, sandwiching an electron-transparent layer. This structure is especially conspicuous after fixation and staining with heavy metals (osmic acid, permanganate, uranyl acetate, etc.). The black layers are then interpreted as proteins adsorbing basic compounds and the clear layer as a stratum of lipids with saturated fatty acid chains that do not reduce osmic acid.

Sometimes the staining of the two dark layers is unequal, indicating that the cytomembranes are not symmetrically equilibrated but have a polar structure (Fig. 2b). This must be expected because in contrast to passive permeability, active translocation proceeds in a polar manner. Thus the outer layer in the unit membrane of the plasmalemma is often less densely stained than the inner one. Robertson (1965) changed his indefensible symmetrical model of 1959 six years later into a polar one by assuming that the outer layer represents not protein but the mucopolysaccharides found in the cell membrane (Fig. 1d).

However, freeze etching (Moor and Mühlethaler 1963, Moor 1964) shows a more complicated structure since the plasmalemma appears to be associated with globular particles of different sizes. In algae they measure 8 nm (Plate II, Fig. C_{1-3}) and in yeast 18 nm (Plate I, Fig. A_{1-2}). The plasmalemma of yeast displays other interesting features: it shows folds which guarantee a 50% surface increase (Plate I, Fig. A_1). Such folds occur also in fungal spores so that they cannot be related to an intensified metabolism. The fact that they are more frequent in resting than in metabolizing cells raises the question whether they may be caused by a shrinkage of the protoplast due to dehydration.

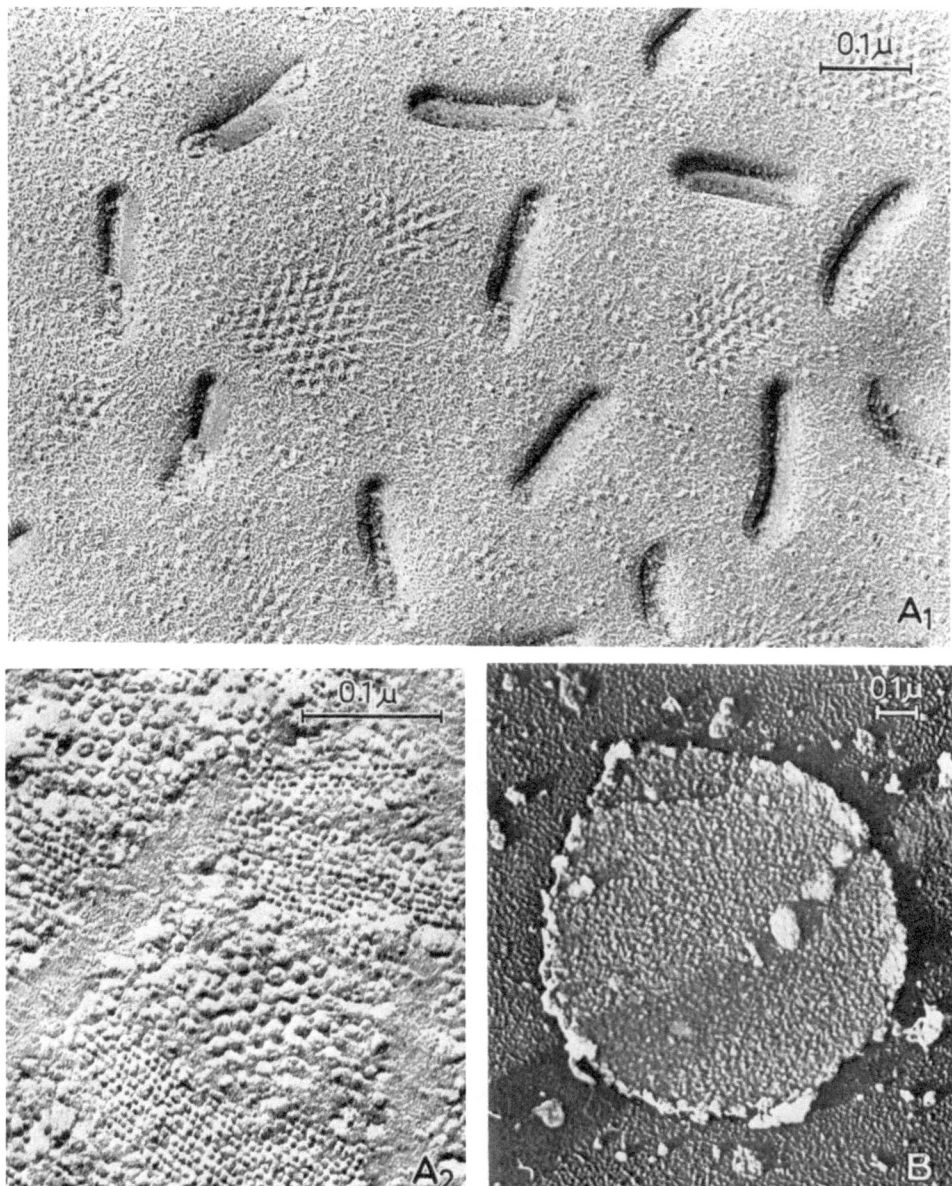

Plate I. Fig. *A*. Freeze-etched plasmalemma of yeast cells, inner fracture face (courtesy of H. MOOR); A_1 showing folds and internal 18 nm particles, ×115,000; A_2 the 18 nm particles consist of subunits forming a central hole, smaller particles in regular array, the folds are free of particles, ×180,000. Fig. *B*. First evidence of globular structural elements in thylakoid membranes of chloroplasts, ×50,000 (FREY-WYSSLING and STEINMANN 1953).

In Fig. A_2 the folds appear flattened and a transitory population of 8 nm particles which is characteristic for starved yeast can be observed. It is a reasonable assumption that such globular particles represent enzyme complexes. Yet the specific functions attributed to them are still speculative. In the green alga *Chlorella* they seem to produce elementary fibrils (Plate II, Fig. C_1) of cellulose (STAEHELIN 1966), and in the blue-green alga *Cyanidium,* they gather at the place where bars are produced by local thickening of the cell wall (Plate II, Fig. C_3). In the naked autospore of *Oocystis apiculata* they are aligned in the direction of the future cellulose fibrils (ROBINSON and PRESTON 1972). In the protozoon *Acanthamoeba* the particles measure 9–12 nm; they appear when this naked organism sporulates and produces a cellulosic cell wall. After the formation of the cellulose fibrils their number is considerably reduced (BAUER 1967). In yeast the 18 nm particles seem to produce glucan fibrils (Plate I, Fig. A_1); however, unexpectedly these particles contain no glucose when isolated but only mannose (MATILE *et al.* 1967) which carbohydrate is not found in the fibrils but in the amorphous matrix of the cell wall.

Unfortunately final information on the ultrastructural status of the globular components of the plasmalemma is not yet available. They are found on either side of the membrane and show different sizes (e.g. 8.5 nm and 12–17.5 nm in *Oocystis*). How these particles are rooted in the underlying film is unknown. Therefore, the better known structure of the membrane of the flat vesicles called *thylakoids* in the chloroplasts will briefly be mentioned. As these membranes derive from the plasmalemma (see 1.5.2.) similar ultrastructural principles may be expected.

The particulate nature of the thylakoid membrane was discovered already in 1953, several years before the unit membrane which does not include globular elements was postulated (Fig. 2a). Based on electron micrographs as in Plate I, Fig. *B,* the symmetrical sandwich model with a smooth surface (Figs. 1c and 1d) was replaced by the concept of a membrane with polar block units and an embossed surface (Fig. 2b) which was thought to be a reason for the weak dichroism of the incorporated chlorophyll. PARK and PON (1961) confirmed the presence of globular elements, then called quantasomes, but only the freeze-etching technique (MOOR *et al.* 1961) allowed successful ultrastructural analysis.

Freeze-fractured thylakoid membranes display two types of particle populations. Certain faces show multienzyme complexes of 12 nm diameter in regular array each of which consists of four 6 nm subunits. Other faces are characterized by disseminated 6 nm protein particles (MÜHLETHALER *et al.* 1965). Originally it was thought that these two faces represented the outer (large composite particles) and the inner (6 nm particles) surface of the thylakoid membrane respectively. However, it could be shown that the particle patterns discovered are internal structures of the membrane which splits along the inner non-polar lipidic faces when the frozen chloroplast preparation is fractured at — 100 °C (BRANTON 1966). The images of the two faces obtained after splitting the membrane must fit together; each of them must behave as the negative of the other considered as positive (WEHRLI

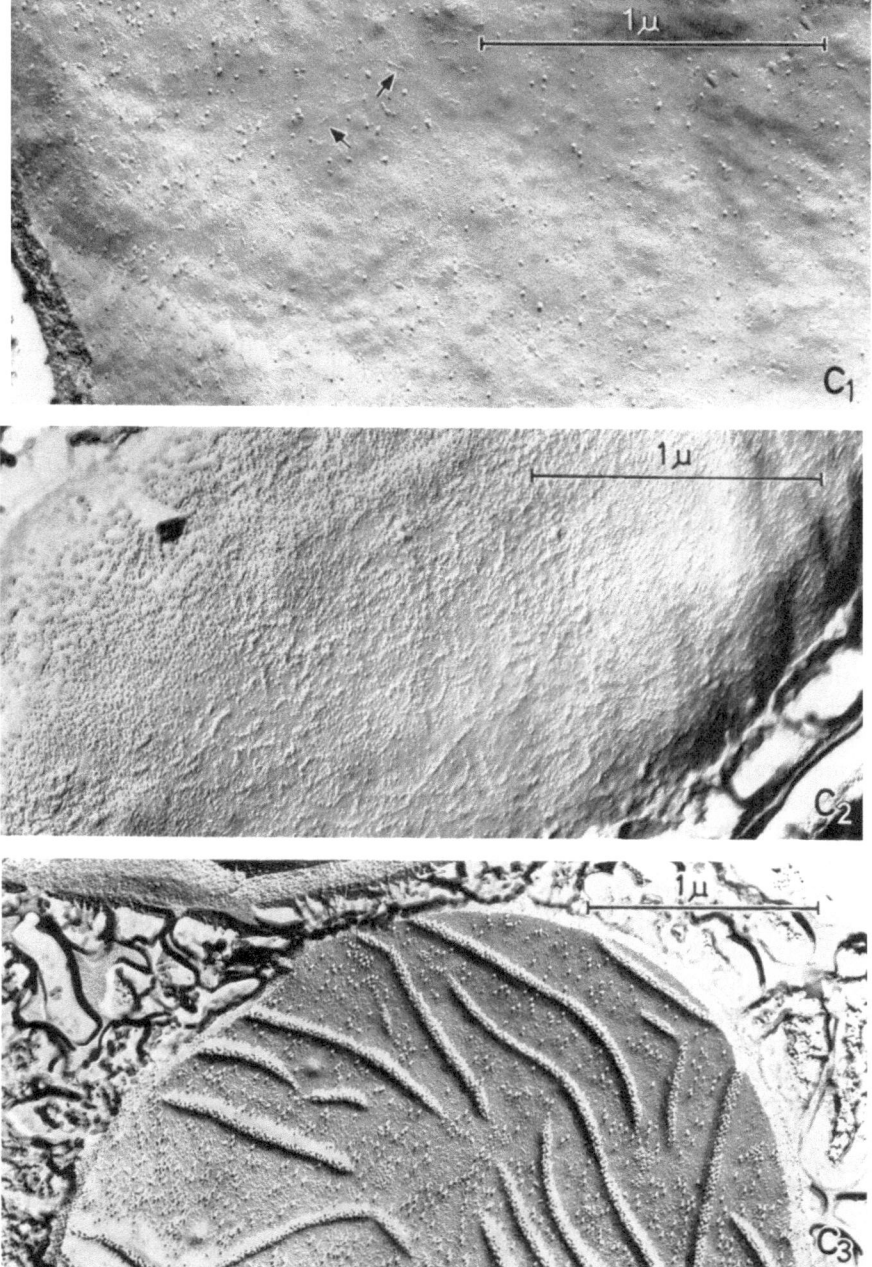

Plate II. Fig. C. Particle populations of the plasmalemma (PL); C_1 and C_2 PL of the green alga *Chlorella* [courtesy of A. STAEHELIN, from Z. Zellforsch. **74**, 325 (1966)]; C_1 particles produce filaments of the size of cellulose elementary fibrils (arrows), ×45,000; C_2 arrangement of particles in linear array, ×38,000; C_3 PL of the blue-green alga *Cyanidium*, agglomeration of particles where cell wall bars are produced, ×30,000 (courtesy of K. MÜHLETHALER).

et al. 1970). Applying this method the ultrastructure of the thylakoid membrane has been disclosed as shown in Fig. 2c (MÜHLETHALER 1971).

According to these findings there are four different faces of the membrane which can be shown: Its outer (OS) and its inner (IS) surface (Fig. 2 c) is uncovered by deep-etching, i.e. removing all the ice above or below the membrane by sublimation, whereas fracturing reveals the outer (OFF) or the inner (IFF) fracture faces. Fig. 2 c shows where the different protein particles of the thylakoid membrane are located. Whilst there are globular macromolecules also on the surface, the bulk of them is situated inside the membrane! When it is split, the smaller particle type adheres to the outer side while the large 12 nm blocks composed of 6 nm subunits are embedded in and tower above the inner side. The internal protein particles prevent the cytomembranes from shrinking when their lipids are removed by extraction.

A similar ultrastructure has been disclosed in the isolated membrane of the vacuole (tonoplast) of yeast cells (KOPP 1971). Deep-etching reveals a rather smooth surface (a in Plate VIII, Fig. S), whereas the fractured membrane displays a population of particles (b in Fig. S). In certain places the particles are lacking; here the inner fracture face appears to be smooth (c in Fig. S).

These facts seem to indicate that the central layer of the cytomembranes consists of proteins (MÜHLETHALER 1972). As to the polar lipids, X-ray analysis shows that bimolecular layers, as established in myelin, occur also in the plasmalemma. However, their polar groups are not covered by proteins as in the models of Figs. 1c and d. In ghosts of red blood cells, the polar groups can be removed by phospholipase C without disintegrating the membrane or changing the conformation of its proteins as determined by ultra-violet circular dichroism (LENARD and SINGER 1968). Therefore the polar groups must form the outer surface of the membrane. However, it is not yet agreed upon what the topical relations between the central protein and the bilayered lipids are. As the plasmalemma holds about equal amounts of proteins and lipids (see Table 1, p. 23), the solution of this problem is important. The tentative model published by MEYER and WINKELMANN (1970) seems too simple and too static to explain the highly dynamic character of the plasmalemma.

A more dynamic modell which includes the lipids has been proposed by KOPP (1972). It is based on double replica evidence from the PL of yeast and is reproduced in Fig. 2 d. The protein molecules between the lipid layers may change their position (arrows). At certain places there are lipidic double layers without protein which yield a bilayer X-ray pattern.

1.2. Dynamics

The stable bimolecular film concept (Fig. 1c) is not only unlikely on account of its lacking polarity but above all because it is essentially static. Of course, this reproach is valid for every membrane model, but the sandwich model is by definition an equilibrium structure, while models with globular or block units (Fig. 2b) are not based on such an assumption.

As biomembranes display their properties of semipermeability and of active ion transport only in the living state, it is obvious that these actions are bound to and correlated with metabolic activities. The chemical reactions prevent a static equilibrium of the membrane structure. It must be in a labile state, the tendency of which to reach a stable state is constantly counteracted by a continuous investment of energy.

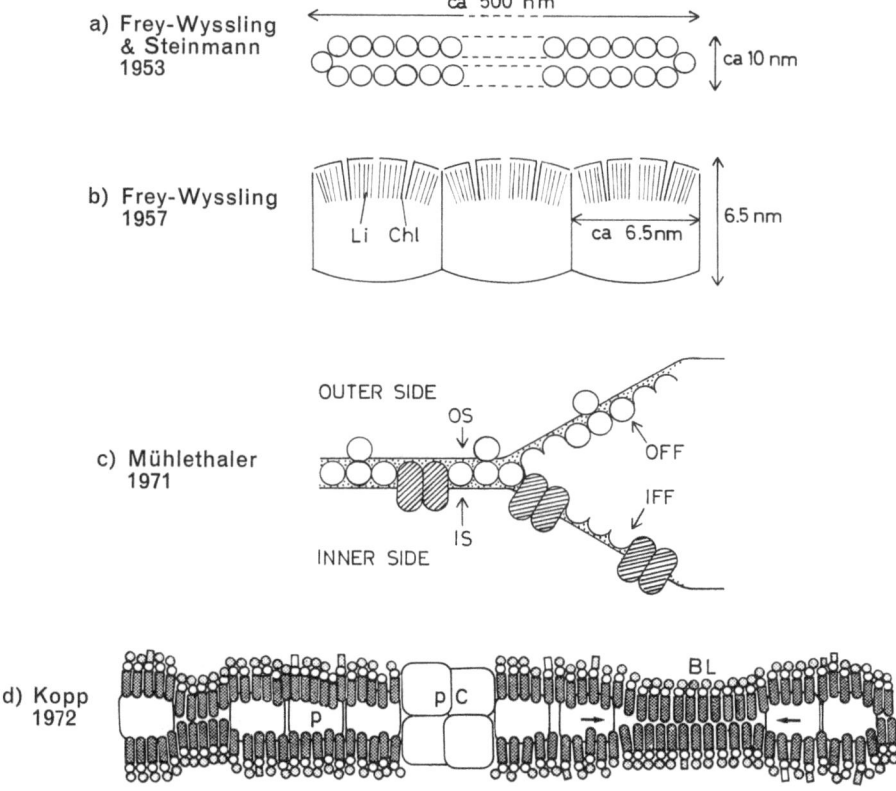

Fig. 2. Globular structural elements of biomembranes. *Li* lipid, *OS* outer surface, *IS* inner surface, *Chl* chlorophyll, *OFF* outer fracture face, *IFF* inner fracture face, *P* protein, *PC* protein complex, *BL* lipidic bilayer.

For the selection and the active transport of ions (SOLLNER 1970, WIPF and SIMON 1970), special carriers have been found which convey for instance potassium ions through a membrane. Such carrier molecules can change their conformation and so trap K^+ ions which are incorporated inside the molecule according to their coordination number 8 (Fig. 3). The available space does not allow the incorporation of Na^+ or other ions so that the well known selectivity of ion intake results.

How the trapped K^+ ion passes through the membrane is another problem (WIPF *et al.* 1970). In artificial membranes the translocation occurs in the following way: The outside of the carrier molecule is lipophilic (Fig. 3), so that it can migrate through an artificial lipophilic film and if an electric potential is administered between the two sides of the membrane, the carrier will diffuse through the film. As such potentials exist also in the plasmalemma, the selective transport of K^+ ions into the living cell seems to be explained. However, for such considerations it must be remembered that in physiology electric potentials are never the cause but on the contrary the consequence of metabolic activities. So it is not known which biochemical reactions produce the necessary potentials and whether they have the same effect in living biomembranes as in artificial equilibrium films on which they are induced.

In the plasmalemma the carrier function seems to be realized by protein molecules which operate as transport enzymes. In bacteria two types of such *permeases* have been found (PARDEE 1968, SITTE 1969 b, p. 28). There is a class of relatively small transport proteins (molecular weight around 3.10^4) which can be isolated from the plasmalemma and crystallized. These have specific binding affinities to certain ions or micromolecules such as SO_4^{2-}, Ca^{2+}, leucine, galactose, etc. A second type of large molecular size deserves special attention because it is related to the energy donor system ATP. These macromolecular transport proteins are anchored in the plasmalemma and cannot be isolated without destroying the membrane. For instance there is an energy-supplying protein HPr which is involved in the transport of sugars through the bacterial cell membrane. It is phosphorylated to P-HPr and has the necessary energy for the enzymatic transphosphorylation

$$\text{P-HPr} + \text{sugar} \xrightarrow{\text{Enzyme}} \text{sugar-6-P} + \text{HPr}.$$

The phosphorylated sugar is released into the groundplasm from which it cannot escape back through the plasmalemma which is impermeable to sugar phosphates (SIMONI et al. 1967).

It is proved that ATP metabolism is the energy source for living membranes. However, whether the energy released operates by means of electric potentials, redox potentials, activated diffusion potentials, or by some other means is not clear.

Morphologically the dynamics of these processes are visible in regions where maximum transport, mitosis, and/or growth occur. In every such case the electron microscope shows a highly sinuated plasmalemma with conspicuous invaginations and dramatic exocytosis processes (FREY-WYSSLING 1962). However, in photographs like Plate III, Fig. *E*, it is difficult to decide whether fixation artifacts have contributed to the irregularly convoluted aspect of the plasmalemma. These doubts can be dispelled by Plate III, Fig. *D*. This electron micrograph (ALBERSHEIM 1965) shows that, after mitosis, only the plasmalemma along the growing septal wall displays the highly dynamic state described, while along the established cell walls the membrane is smooth without any sign of highly active metabolism.

In this way it is demonstrated how over short periods the plasmalemma must be active in a way which revolutionizes its morphology by convulsive movements.

The folds of the plasmalemma mentioned earlier are another sign of its surface dynamics. Size and number of the folds vary during the life time of a cell. Yet, contrary to all expectations they are less numerous in metabolizing than in resting cells, especially in fungal spores where they appear in large numbers or may even abound. Such observations have raised the question whether the folds are perhaps fixation artifacts. However, this criticism can easily be eliminated because freeze-etching furnishes true to life pictures (Plate I, Fig. A).

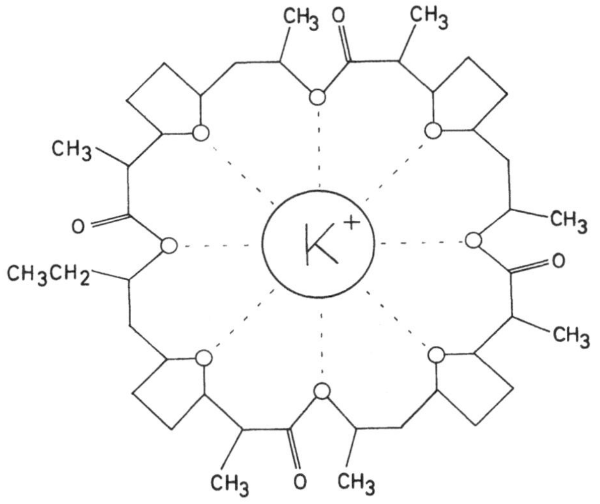

Fig. 3. Projection of the molecule monactin with selective permeability for potassium ions.

1.3. Growth

The surface of extending plant cells in filaments of grass stamens can increase up to 65 times per hour (FREY-WYSSLING 1948), i.e. every minute the plasmalemma must be enlarged by its original size. This rapid growth occurs by intussusception which seems to require the synthesis and intercalation of new molecules between the existing ones without increasing the width of the membrane. As long as the plasmalemma was believed to be a bimolecular lipid layer, such a process was easily conceivable in that lipid molecules of the cytoplasm could be considered as reserve material for intercalation. However, the ultrastructure of the plasmalemma is so complicated that it is difficult to visualize a simple mechanism of molecular synthesis inside the membrane which would guarantee rapid surface enlargement without interfering with the functions of the membrane. These concomitant physiological activities embody the riddle of growth.

If a naked protoplast is cut into pieces by microsurgery, the liquid groundplasm flows out and is no longer covered by a plasma membrane. However, after a short time the new surface shaped by surface tension produces a new plasmalemma which coats the plasmic droplet.

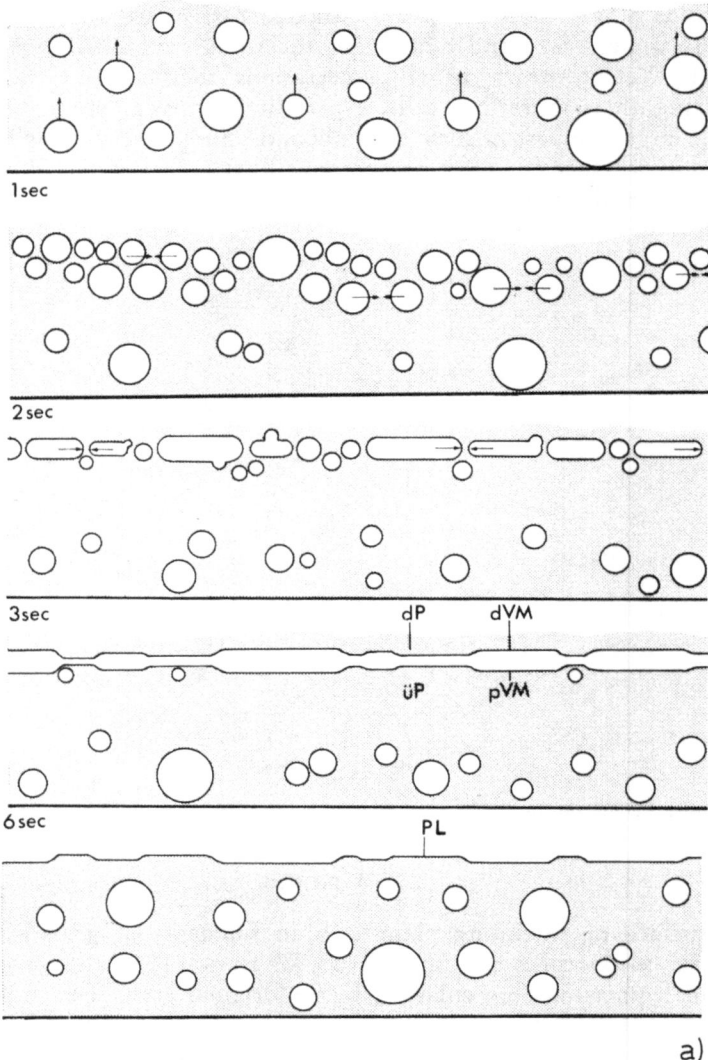

a)

Fig. 4. *a*) Regeneration of the PL on an uncovered plasmic surface of the slime mould *Physarum polycephalum. dP* degenerating plasm, *üP* surviving plasm, *dVM* distal, *pVM* proximal part of the fused vesicular membranes [courtesy of K. E. Wohlfarth-Bottermann, from Wilh. Roux' Arch. **164**, 321 (1970)].

Therefore, the plasmalemma must be regenerated in a short time. How this rapid healing is realized has been demonstrated by Wohlfarth-Bottermann and Stockem (1970) with the slime mould *Physarum polycephalum*. The plasmodium is punctured with glass capillaries and the plasmic cylinder obtained fixed for inspection in the electron microscope immediately (zero seconds) or after 1, 2, 3, or 6 seconds (Fig. 4*a*). It is observed that vesicles containing slime which are dispersed in the groundplasm gather under the uncovered periphery and fuse, forming a flat vacuole parallel to the surface. The plasmic margin outside the vacuole and the distal vacuolar membrane degenerate, while the proximal vacuolar membrane represents the plasmalemma of the surviving part of plasm. Therefore the plasmalemma has been regenerated by the fusion of the membranes of slime vesicles which probably represent Golgi vesicles. Most amazing is the rapidity of the processes involved: Immediately after the loss of the original plasmalemma, the vesicles strive to gain the uncovered surface (Fig. 4a). Their number and total surface

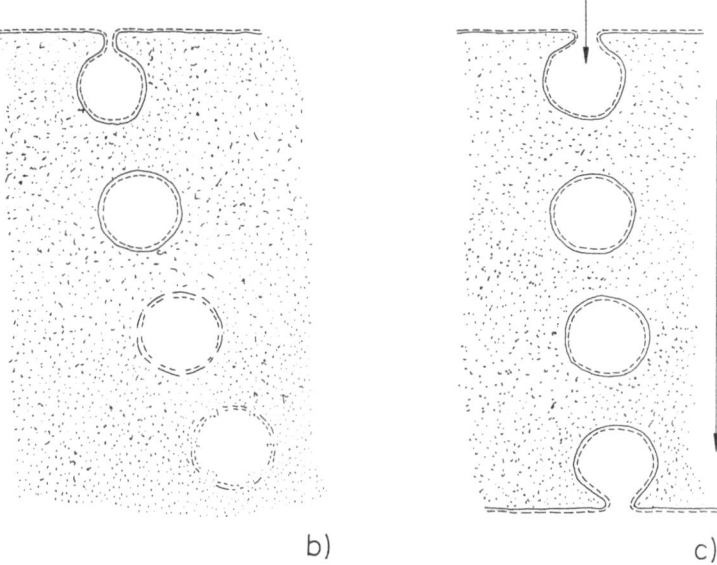

b)   c)

Fig. 4. b) Endocytosis. The vesicular PL membrane is digested. c) Cytopempsis. The endocytotic vesicle is secreted on the opposite side of the cell (cp. Plate III, Fig. F) (A. FREY-WYSSLING and K. MÜHLETHALER, from Ultrastructural Plant Cytology, Elsevier 1965).

seem to increase (by dispersion?). After 1 second fusion starts, after 2 seconds the fused vesicles flatten and already after 3 seconds the formation of the extended vacuole with the new plasmalemma is accomplished.

The same means of using prefabricated plasmalemma around membrane bounded Golgi vesicles is applied for the extension growth of the cell membrane mentioned above. Thus, for rapid growth or regeneration of the plasmalemma intussusception is not achieved by intercalation of individual molecules but by fusion of preformed patches of plasmalemma with that already existing, and so expanding its surface (see 2.3.3.2.).

The final question is how the Golgi apparatus with its complicated and highly specialized ultrastructure is able to synthesize plasmalemma. Whether it works according to the model of preexisting structures in its own membrane, or whether it is capable of forming plasmalemma from micromolecules in the groundplasm or from structural elements left over from the membranes of endocytotic vesicles which have disappeared, is not clear so far.

1.4. Functions

Since the animal cell is naked and the plant cell wall most often holopermeable, the plasmalemma is the organelle which has to contend on the inside with cellular and on the outside with extracellular conditions. As a consequence the functions of this barrier are manifold.

1.4.1. Resorption

Any substance which is brought into the cell must pass through the plasmalemma. In the case of water this intake is achieved by osmosis, a process made possible by the semipermeability of the cell membrane. As already mentioned, it functions passively as long as the labile ultrastructure of the plasmalemma is maintained by energy supply and disappears when the necessary ATP regeneration by photosynthesis, respiration or fermentation becomes inoperative. An active intake of water through the plasmalemma has not been observed so far. However, as shown below, an active pumping of water into the cell can occur by *pinocytosis* (Greek *pinein* = to drink) whereby the plasmalemma displays local growth in the form of invaginations (LEWIS 1931, HOLTER 1959, MAYO and COCKING 1969).

Intake of ions and micromolecules (sugars, aminoacids) can occur by diffusion only "downhill", i.e. if the concentration is lower inside the plasmalemma. However, resorption works mostly "uphill", especially in the case of inorganic ions the concentration of which cannot be reduced inside the cell by polymerization or metabolism. In this case an active transfer is necessary by means of "permeases" or "carriers" (Fig. 3) for which energy must be expended.

Ultrastructural particles and macromolecules which are difficult to translocate through the plasmalemma (e.g. ferritin, proteins, ribonuclease (BRANDT and PAPPAS 1960, FAWCETT 1965) can be incorporated into the cell by *endocytosis*. This term was chosen instead of pinocytosis when it became evident that not only water but also dissolved (FREY-WYSSLING 1963) and solid substances can be brought into the cell by local inward folds of the plasmalemma. These invaginations detach from the cell surface, forming plasmalemma-bound vesicles which wander centripetally into the cortical groundplasm.

The destiny of these vesicles varies. They may remain intact in the plasm for some time or their membrane may be digested and disappear whereupon their contents are incorporated in the groundplasm (Fig. 4 *b*). This process must be highly dynamic (Plate III, Fig. *E*).

A third possibility is that the vesicles migrate across the whole cell and their contents are then extruded on the opposite side. Such behavior is characteristic for the epithelial cells of the blood capillaries (Plate III, Fig. *F*). In this case the plasmalemma is involved not only in resorption but also in translocation phenomena. When the vesicles touch the opposite surface, their membranes fuse with the plasmalemma and an opening is formed. This process is called *exocytosis* and the translocation as a whole was termed *cytopempsis* by WOHLFARTH-BOTTERMANN (1963).

These observations raise interesting questions. By which principle is the formation of plasmalemma-bound vesicles governed? How is the necessary energy invested in these processes? Similar contractile structures seem to be responsible for the movement of vesicles inside the groundplasm as those which are involved in the dynamics of plasmic flow (WOHLFARTH-BOTTER-MANN 1968). In amoebae and acellular slime moulds the fluid endoplasm is moved by a contractile exoplasm. In its state of active contraction it is characterized by plasmic filaments of 4 to 21 nm width (WOHLFARTH-BOTTER-MANN 1967). These filaments persist after a prolonged extraction with glycerol and contract with ATP in analogy to acto-myosin of muscles (KAMIYA and KURODA 1965). Proteins similar to actin (F-actin) and myosin (myxomyosin) have been isolated from the plasmodia of the slime mould *Physarum polycephalum*. The 6 nm broad filaments of F-actin disintegrate into 6 nm globular protein molecules on the addition of a 2 millimolar $MgCl_2$ solution. It is an attractive hypothesis to consider the stiffening contraction and the liquefying relaxation of the filamentous system as a reversible polymerization and depolymerization of the 6 nm globular protein units.

Morphologically of great interest is the detachment of a part of the plasma-lemma which persists during cytopempsis and can be re-fused with another part of the PL (Fig. 4 *c*). On the other hand its possible digestion by the groundplasm is of first importance (Fig. 4 *b*). This means that the molecular elements of the plasmalemma must be present in the groundplasm, and the question is whether it can not only break down plasmalemma but also resynthesize it from the dissolved elements. This would involve the possibility of a *de novo* formation of the plasmalemma by the groundplasm.

1.4.2. Elimination

Exocytosis permits all kinds of elimination processes: extrusion of assimi-lates (secretion), of dissimilates (excretion), and of not metabolizable resorbates (recretion) such as salts, e.g. NaCl (FREY-WYSSLING 1935).

Plate III, Fig. *G*, shows how terpenes are eliminated from a plant cell (SCHNEPF 1965). They originate in droplets which are coated with a mem-brane. When the contents of these vesicles are eliminated by exocytosis, the vesicular membranes fuse with the plasmalemma. This is an indication of the identity of these two membranes and therefore, the problem arises whether the membranes of the terpene vesicles are produced by the groundplasm (proving its capacity of forming a *de novo* plasmalemma) or whether they represent Golgi vesicles (see 2.4.) or ER derivatives.

1.4.3. Protection

The wall which functions as a protecting skin and exoskeleton of the plant cell is produced by the activities of the plasmalemma. The wall consists of two components: an amorphous plastic gel matrix with a high water content and a reinforcing fibrillar system which gives the wall its stability and elasticity as well as its optical (birefringence) and mechanical anisotropy.

The matrix is composed of hemicelluloses (especially uronides) and pectic material, the carboxylic groups of which account for the extremely high hydration of young cell walls. The matrix compounds are produced by the Golgi apparatus and secreted through the plasmalemma by exocytosis (see 2.3.3.).

The fibrils of the reinforcing system consist of cellulose (β-1,4-poly-glucosan), glucan (β-1,3-polyglucosan) in yeast or chitin in other fungi. They are synthesized in the matrix outside the plasmalemma. In the case of cellulose, elementary fibrils with 3.5 nm diameter and indefinite length appear. They are crystalline, hold less than 40 straight chains of cellulose molecules and grow by concomitant polymerization and crystallization of β-glucose molecules (FREY-WYSSLING 1969). The necessary extracellular polymerase seems to need uridine-diphosphate glucose (UDPG) for the formation of β-1,4-glucosan chains (COLVIN 1964, BEN-HAYYIM a. OHAD 1964). The uridine diphosphate (UDP) freed is transformed into triphosphate (UTP) which energy source helps in regenerating UDPG from glucose-1-phosphate (HASSID et al. 1959):

$$UTP + \alpha\text{-D-glucose-1-phosphate} \leftrightharpoons UDPG + \text{pyrophosphate}.$$

Although it is not yet possible to produce elementary fibrils *in vitro* with the substrates mentioned above and the necessary enzymes, biosynthesis of cellulose fibrils appears to proceed along such lines. It is a plausible hypothesis that certain protein particles of the plasmalemma represent enzyme complexes for this purpose, and their connection with filamentous (Plate I, Fig. A_1) or fibrillar (Plate II, Fig. C_1) structures on electron micrographs seem to support such speculations. However, with isolated particles no biochemical or enzymological proof can be forwarded as yet for this view (MATILE et al. 1967, FRANZ and MEIER 1969).

Notwithstanding that the mode of formation of the fibrils and of their arrangement to form dispersed, tubular or parallel helical textures are contro-versial (FREY-WYSSLING 1962), it is a fact that these morphogenetic mani-festations of cellular life occur outside the plasmalemma. In contrast to the reinforcing system, the matrix of the cell wall is produced inside the cell and behaves passively without special morphogenetic aptitudes after having been secreted beyond the plasmalemma.

1.5. Derivatives of the PL Membrane

Some of the manifold functions of the plasmalemma, whereby the composi-tion and ultrastructure of the membrane undergo alterations may be emphasized. An important example of such a metamorphic transformation is the formation of the myelin sheath of nerves.

1.5.1. Myelin Sheath

Nerves must be highly insulated because nervous conduction is based on the transmission of electric impulses. For this purpose a highly lipophilic sheath is wrapped round the conducting axon of the nerve. This medullary

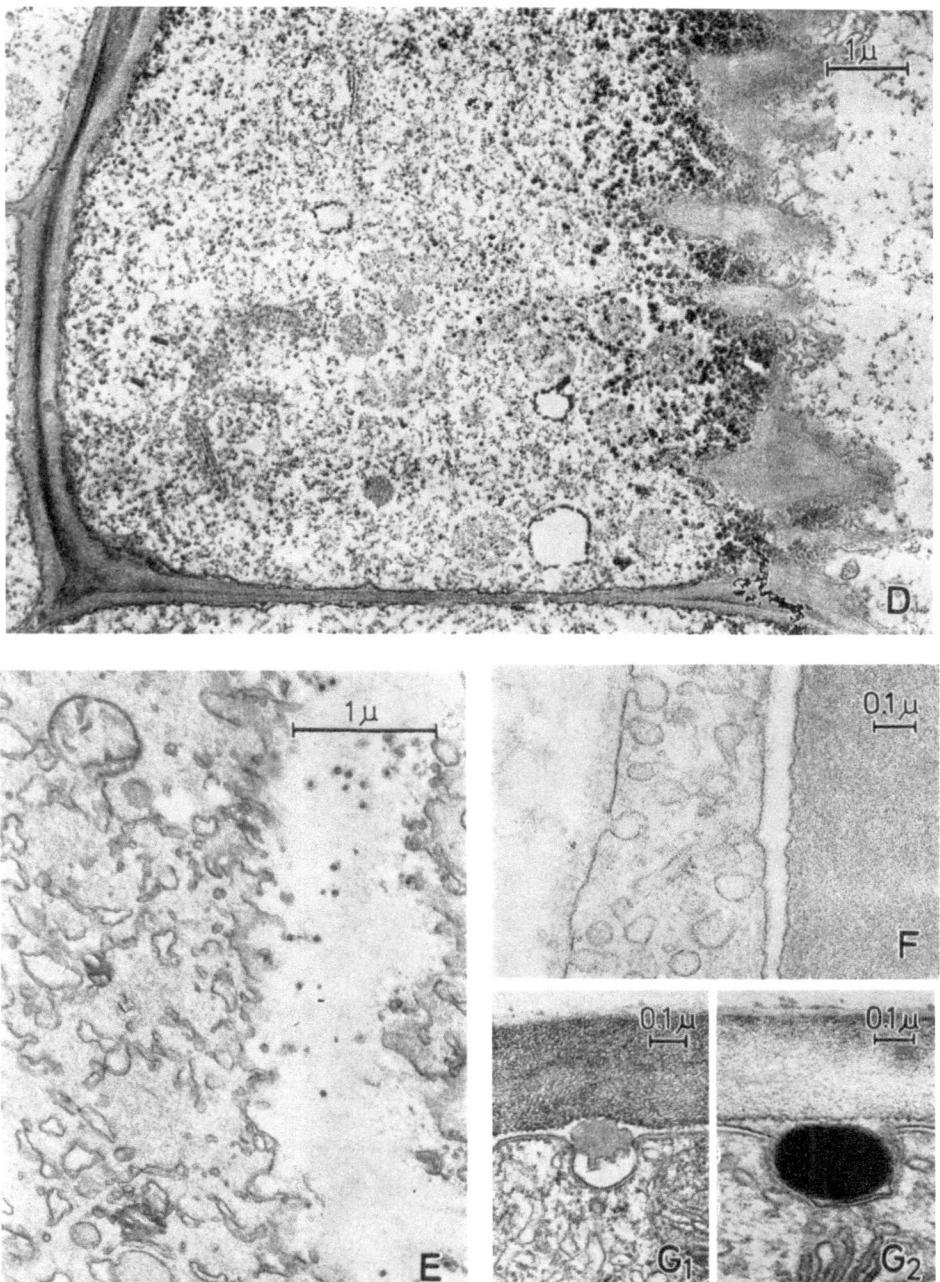

Plate III. Fig. *D*. Dynamics of the PL during the formation of a new cell wall, root tip of onion, Golgi vesicles stained by hydroxylamine and iron [courtesy of A. ALBERSHEIM, from Protoplasma **60**, 131 (1965)], ×10,000. Fig. *E*. Dynamics of the PL during growth of the cell wall in width, root cortex of Ricinus, ×20,000 (courtesy of K. MÜHLETHALER). Fig. *F*. Cytopempsis of PL bound vesicles through the endothelian cell of a blood capillary, ×50,000 (courtesy of A. VOGEL). Fig. *G*. Exocytosis of terpenoids in the spadix-appendix of the aron lily *Typhonium divaricatum*, ×50,000 [courtesy of E. SCHNEPF, from Planta (Berlin) **66**, 374 (1965)].

sheath originates from a spiral pseudopodial growth of so-called Schwann cells around the neural axon (Fig. 5). The nucleus of each such glial cell is situated behind the tip of the pseudopodium and the cytoplasm is withdrawn from the inner convolutions. In this way a spiral is formed consisting entirely of plasmalemma. As a consequence the neural sheath is a lamellated body with a periodicity of 15 nm. Each period holds two 7.5 nm wide cell mem-

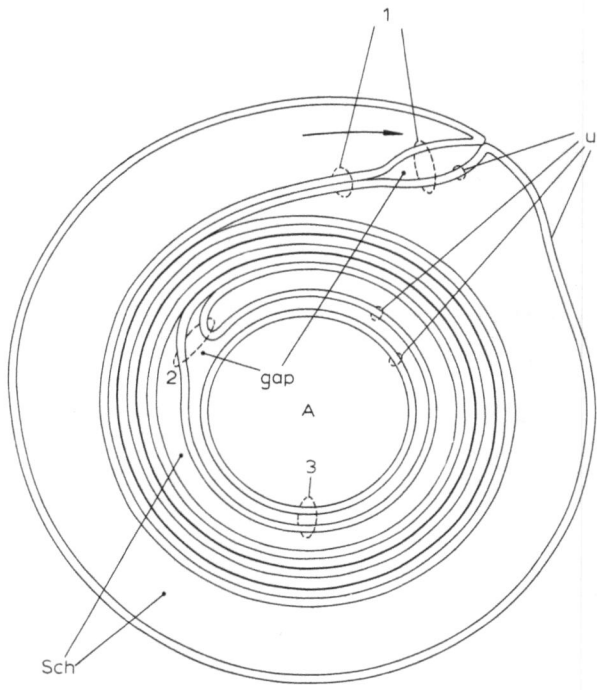

Fig. 5. Ontogeny of the nerve myelin sheath by spiral growth of the Schwann cell (*Sch*). → direction of the pseudopodial growth. *A* neural axon, *u* unit membrane, *1, 2, 3* double membranes of different types [J. D. ROBERTSON, from Biochem. Soc. Symposia **16**, 3 (1959), Cambridge University Press, Mass. (USA)] (cp. Plate IV, Fig. *H*).

branes pressed against each other. They are enriched with lipids, the arrangement of which can be disclosed by X-ray analysis (FINEAN 1953) and electron microscopy (FLUCK *et al.* 1969).

Whilst the plasmalemma holds anything from equal amounts of protein and total lipids up to a protein to lipid ratio of 4, this relation is only 0.25 in myelin (Table 1). Thus myelin is 4 to 16 times richer in lipids than typical plasmalemma. It is evident that such a change in chemical composition must also alter the ultrastructure of the membrane and therefore, the structure of this metamorphosed object was an unsuitable choice as the model for the unit membranes and for biomembranes in general (Plate IV, Fig. *H*).

The active functions of resorption and elimination are lost in the myelin sheath, and the membranes behave merely as a stack of passive insulating layers. Consequently their ultrastructure must be simple as compared to that of a plasmalemma in full action with metabolism and energy transfer. Therefore, the claim that the unit membrane structure should be that of the myelin lamellae (ROBERTSON 1960) must be abandoned. In physiologically active cytomembranes with 4 to 16 times more protein than in myelin, the lipids can scarcely have the same structural importance as in the sheaths of nerves.

Table 1. *Protein and Lipids in Cytomembranes*
Lit. (1) MATILE et al. 1967, (2) KORN 1969, (3) SITTE 1969.
PL plasmalemma, ER endoplasmic reticulum, Mi mitochondria

		$\dfrac{\text{Protein wt}}{\text{Lipid wt}}$
PL	Intestinal villi (2) (3)	4.6–2.3
	Erythrocytes (2)	4.0–1.5
	Liver cells (2) (3)	2.3–1.0
	Yeast (1)	1.05
	Thylakoids of chloroplasts (2) (3)	0.96–0.80
	Myelin of nerve sheath (2) (3)	0.25
ER	Endoplasmic reticulum (2)	6.7–0.9
	Nuclear envelope (3)	4.1
	Microsomes (3)	2.3
Mi	Inner membrane (2) (3)	3.6–3.3
	Outer membrane (2) (3)	1.2–0.95

1.5.2. Thylakoids

In the primitive prokaryotic cells (bacteria and blue-green algae), the plasmalemma performs additional functions which in eukaryotes are carried out by mitochondria or plastids. In certain bacteria NADH and ATPase of the respiratory chain have been found in the cytoplasmic membrane (GOSH and MURRAY 1969), so that in the absence of mitochondria it seems to function as the organelle for respiration. This fact led to the theory that the mitochondria of the eukaryotes represent such bacteria which have been incorporated in the cells as symbionts at an early stage of evolution (COHEN 1970, NASS 1971). Another hypothesis would be to assume that the inner respiratory membrane of the mitochondria derived at some time from the plasmalemma, possibly by a transformation of the inner membrane of plastids which are plasmalemma-born.

The hypothesis of symbiosis presupposes a heterotrophic fermentating amoeboid cell which is invaded by autotrophic algae and respiring bacteria, i.e. it assumes a *polyphyletic* origin of life. It is true that in their valuable discussion concerning the pro and contra of this hypothesis, SCHNEPF and BROWN (1971, Fig. 12) introduce a monophyletic prokaryotic ancestor from which the three types of phagocytizing, of photosynthesizing and/or aerobic-respiring organisms derived. However, if this ancestor had the faculty of producing the three metabolic types necessary for the creation of a eukaryotic plant, it may be asked why an evolution along three divergent lines was necessary.

For a real *monophyletic* development one may argue in the following way: Vital activity could only start after the creation of an energy donor for metabolic processes which received its free energy from some photoreaction. As a consequence, life must have started by a kind of primitive photosynthesis which, even in its highly developed actual state is still characterized by the production of the energy donor ATP. Since at that time of geohistory there was no oxygen available in the atmosphere, no respiring bacteria were possible and, therefore, in a monophyletic file the mitochondria could possibly have derived from plastids from which they inherited the ATP system, cytochrome, NAD (in the form of NADP), cardiolipin and other special agents.

According to the theory of symbionts the two membranes of the plastidal and the mitochondrial envelope would be plasmalemma membranes. Incorporated blue green algae and bacteria would have brought along their plasmalemma, while the outer membrane of the organelles would represent the boundary of endocytotic vesicles detached from the plasmalemma of the host. The established differences between the inner and the outer membrane of the envelope would then be due to the original differences of the plasmalemma of the symbiont and that of the host (e.g. that the latter was not capable of producing thylakoids).

In a monophyletic view, the chloroplasts must have originated by an encirclement of plasmalemma-born thylakoid stacks, together with inter or circum-thylakoidal groundplasm containing the genetic system found in these organelles, by a process similar to that of the formation of the nuclear envelope (see 3.1.2.) or that described under 1.3. (Fig. 4 *a*), namely by the fusion of a file of coated vesicles. As a consequence the inner and the outer membrane of the resulting envelope should be alike, in the first case of ER and in the second case of PL-nature. Since this is not the case, divergent phylogenetic differentiations of the two membranes must be admitted when postulating that the chloroplasts originated in this way.

The process of the *ontogenetic* development of thylakoids by invaginations of the plasmalemma can be observed in phototrophic prokaryotes. In these organisms no fully developed plastids with a double membrane occur but only the photosynthesizing membranes embedded without any boundary in the groundplasm. As in the chloroplast they represent flat lamellar vesicles which are thus identical with thylakoids. In *Rhodopseudomonas spheroides* these thylakoids are continuous with the plasmalemma (Plate IV, Fig. *K*, Drews and Giesbrecht 1963). In *Rhodopseudomonas viridis* the thylakoids are generated by obvious invaginations of the plasmalemma (Plate IV, Fig. *I*, Giesbrecht and Drews 1966). Similar statements were made for the blue-green algae *Oscillatoria* (Jost 1965) and *Nostoc* (Schnepf 1964). Therefore, it is an established fact that in prokaryotes the thylakoids derive from the plasmalemma. Of course, specializing for the function of photosynthesis, they must undergo certain internal metamorphoses as shown by Drews et al. (1969).

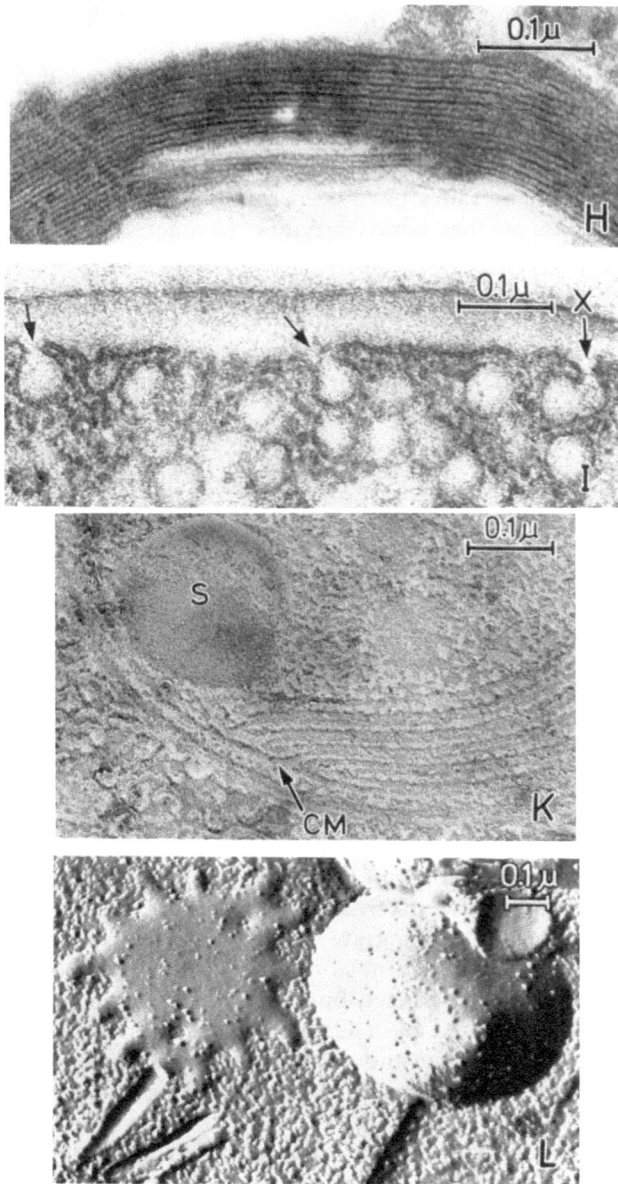

Plate IV. Fig. *H*. Lamination of myelin of human brain, ×150,000 (courtesy of A. VOGEL), compare Fig. 5. Fig. *I*. Formation of thylakoids by invagination of the PL (arrows) in photosynthetic bacteria, ×125,000 [courtesy of G. DREWS and P. GIESBRECHT, from Zentralbl. Bakteriologie **190**, 508 (1963)]. Fig. *K*. Evidence of continuity of thylakoids with PL (cell membrane CM), *S* reserve material, ×115,000 [courtesy of G. DREWS and P. GIESBRECHT, from Arch. Mikrobiol. **54**, 297 (1966)]. Fig. *L*. Freeze-etched Golgi apparatus in the yeast *Schizosaccharomyces*, vesicles at the rim of the cisternae, an extending tubule is broken off, ×50,000 (courtesy of F. KOPP).

So thylakoid and plasmalemma membranes are not identical but phylogenetically and ontogenetically related, i.e. *homologous.*

If the chloroplasts of the eukaryotes derive from the so-called chromoplasm of the phototrophic prokaryotes, their inner membrane from which the thylakoids originate would also be homologous with the plasmalemma. As for the outer membrane of the plastids and the mitochondria, unexpected relations with ER membranes have been found. The outer membrane of the mitochondria contains the same enzymes for activation, elongation and β-oxidation of fatty acids as the ER membrane (Nass 1971). In the algal taxa *Xanthophyceae, Chrysophyceae, Bacillariophyceae, Phaeophyceae* and *Euglenophyceae* two chloroplast envelopes may be found: outside the normal envelope with its two membranes, a set of ER cisternae surrounds the plastid (Massalski and Leedale 1969). These cisternae seem to be involved in the transfer of assimilates.

2. Golgi Apparatus (GA)

2.1. Ultrastructure

The Golgi apparatus consists of a stack of flat wafer-like cisternae and a system of spherical vesicles of different sizes given off by the rim of the cisternae. The stack is called a "dictyosome", this Greek word means a "network". The term was chosen for the structure of coagulated Golgi bodies seen in the light microscope (Golgi 1882/1885). However, in the electron microscope the cisternae appear not reticulate as a whole, only their peripheral region may be fenestrated (Morré *et al.* 1971), especially when they are highly active. In general their shape is more like that of a saucer than of a net. From the rim of the saucer vesicles are beaded off and tubules may be generated (Plate IV, Fig. *L*).

The stacks have a polar organization (Grassé 1957). There is a proximal or forming pole, where new cisternae are produced and a distal or secreting pole, where the cisternae disintegrate into vesicles (Fig. 6). These dynamics are especially conspicuous when not only the cisternal rim evaginates locally for the formation of small vesicles, but when the cisterna inflates as a whole so that a large cavity coated by a Golgi membrane is formed. This striking vesicle leaves the dictyosome and is replaced by a new one. In this way the constant shift of the cisternae from the forming to the secreting pole is evident.

The individual cisternae have a fairly constant distance from each other of 20–25 nm. In preparations fixed with glutaraldehyde a thin electron dense line is found in the middle of the narrow zone (Fig. 6). This blanket consists of evenly spread proteinaceous filaments. It is thought that it may represent a compressed layer of the groundplasm with which it is continuous (Ledbetter and Porter 1970). Should this be the case its fabric-like structure would be representative of a similar three-dimensional fibrillar lattice in the groundplasm.

Franke and Scheer (1972) find that the fenestration is due to pores with diameters grading from 60 nm down to 10 nm ("minipores"). In the center

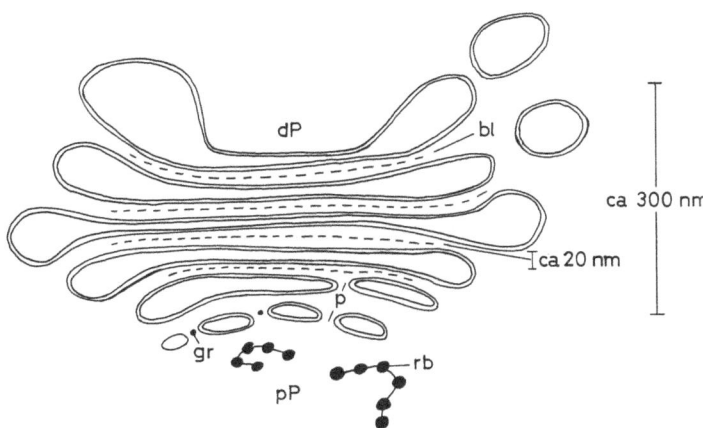

Fig. 6. Idealized dictyosome, drawn according to data from LEDBETTER and PORTER (1970)
and FRANKE and SCHEER (1972). *pP* proximal or forming pole, *dP* distal or secreting pole,
p pores, *gr* granule (of nucleoprotein?), *bl* blanket, *rb* ribosomes.

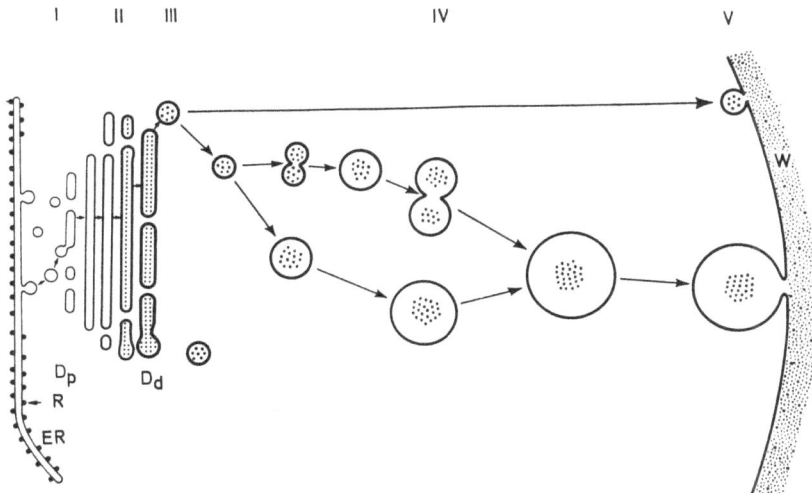

Fig. 7. Functioning of the Golgi apparatus leading to expansion of a fungal hypha at the
apex. I. Material is transferred from ER to the dictyosome by blebbing and refusion of ER
vesicles to form a cisterna at the proximal side of the dictyosome (*Dp*). II. Cisternal contents
and membranes are transformed as the cisterna is displaced to the distal side (*Dd*). III. Cisternae at the distal pole form secretory vesicles. IV. Secretory vesicles migrate to the hyphal
apex. Some may increase in size or fuse with other vesicles. V. Vesicles fuse with the PL of
the apex liberating their contents into the matrix of the cell wall [courtesy of CH. E.
BRACKER, from GROVE *et al.*, Amer. J. Bot. **57**, 245 (1970)].

of the poral canal there is a granule or rodlet similar to the central dot in nuclear pores (Fig. 10). It seems to be in relation with polysomal ribosomes in the groundplasm of the proximal dictyosomal pole (Fig. 6). The number of pores of the proximal cisternae is much larger than that of the mature distal cisternae. Perhaps there is a polar transfer of macromolecules, necessary for the maturation of the secreting cisternae (Fig. 7).

2.2. Ontogeny

It can be shown in lower plants that the Golgi cisternae originate from vesicles of the endoplasmic reticulum (GROVE et al. 1970), or from the nuclear envelope (MOORE and MCALEAR 1963, MASSALSKI and LEEDALE 1969). In both cases the cisternae are formed by the fusion of numerous vesicles given off by the reticulum (Fig. 7) or the outer membrane of the nuclear envelope (Plate V, Fig. M). During the shift of the cisternae from the forming to the secreting pole, there is a pronounced process of maturation. The contents of the endoplasmic reticulum and the perinuclear space which is a dilute watery fluid becomes, as a rule, enriched in monomerous and oligomerous carbohydrates detectable by the periodic acid Schiff reagent (PAS). The membrane which is responsible for this uphill concentration of sugars must undergo a fundamental transformation shown by a considerable increase in thickness. The fact that Golgi membranes (7–10 nm) are substantially thicker than ER membranes (5–6 nm) after osmic fixation (SJÖSTRAND 1963) proves that intussusceptional growth, which may proceed with remarkable speed, takes place. Golgi membranes must be produced at rates of fractions of a minute, since in *Glaucocystis* every 30 seconds a large Golgi vacuole is given off from the stack, yet the number of its cisternae does not diminish (SCHNEPF and KOCH 1966 a).

The origin of the cisternae and their fate at the distal pole are illustrated by a diagram of GROVE et al. (1970) which is reproduced in Fig. 7. The vesicles evaginated at the secreting pole are rich in carbohydrates. They migrate to the cell surface and on the way they may fuse or grow. When they reach the plasmalemma their contents are extruded from the cell by exocytosis. Here we meet with the question: which are the functions of the Golgi apparatus?

Plate V. Fig. *M*. Ontogeny of Golgi cisternae from vesicles evaginated by the nuclear envelope of the Xanthophycean alga *Tribonema*, ×38,000 [cortesy of A. MASSALSKI and G. F. LEEDALE, from Brit. phycol. J. 4, 159 (1969)]. Fig. *N*. Ontogeny of the new cell wall after mitosis in the root tip of *Phalaris canariensis*; N_1 formation of Golgi vesicles and their accumulation in the equatorial plane of the cell; N_2 Golgi vesicles line up, to the right a Golgi apparatus visible in plane view, ER strands prepare future plasmodesmata; N_3 Golgi vesicles fuse laterally forming the cell plate which expands towards the longitudinal walls of the mother cell; N_4 differentiated cell wall with middle lamella and adjacent primary wall layers, N_5 longitudinal wall to which the new wall is attached, N_6 Golgi vesicles showing electron-dense central core; N_7 the dense cores are the precursors of the middle lamella; N_8 growth in width of the cell wall by incorporation of additional Golgi vesicles, $N_{1,2,4,6}$, ca. ×20,000, N_3 ×25,000, N_5 ×7,000, $N_{7,8}$ ×40,000.

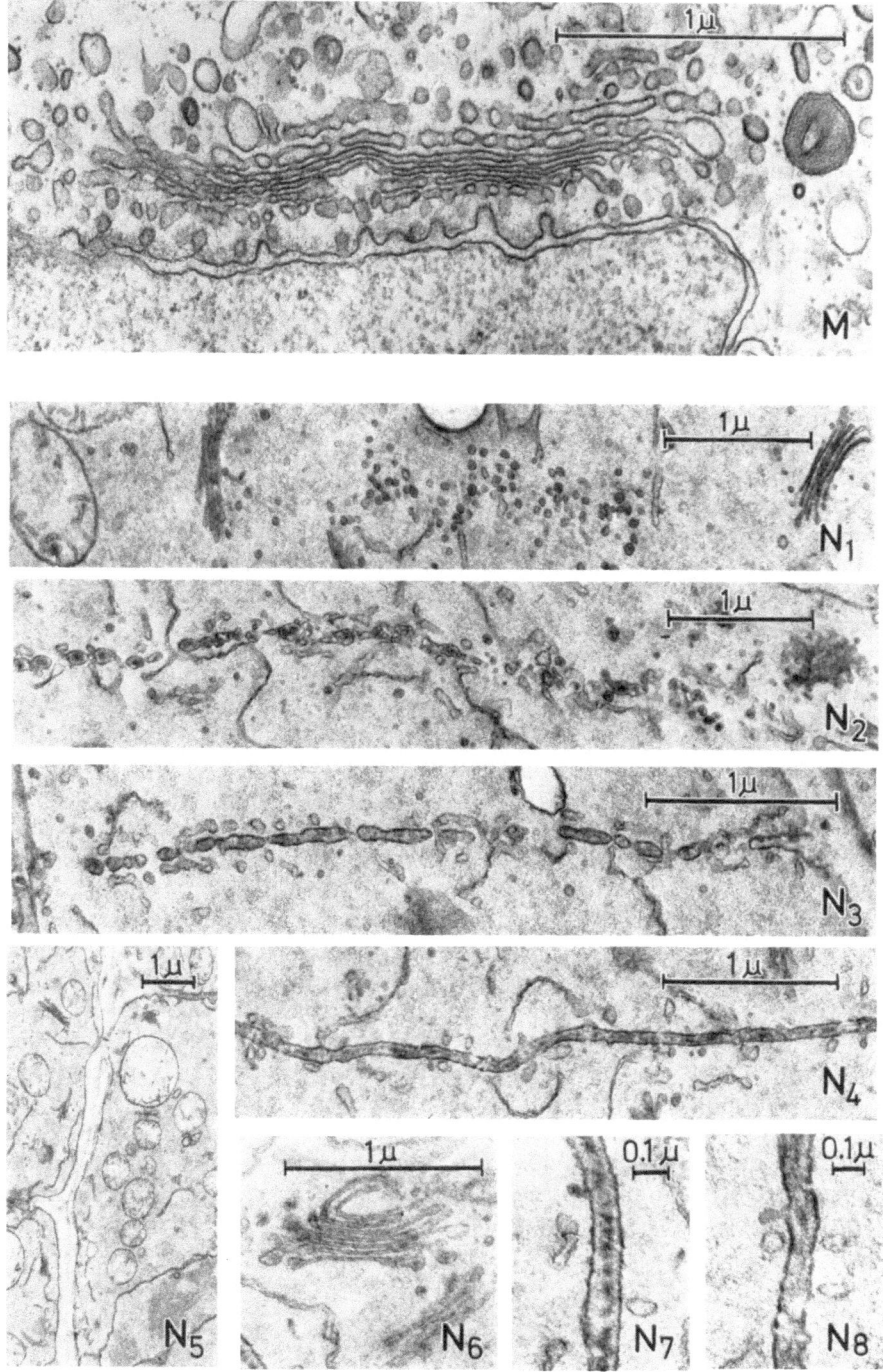

Plate V.

2.3. Functions

2.3.1. Regulation of the Cellular Water Content

In algal cells with "contractile" vacuoles (autospores of *Glaucocystis*, *Vacuolaria virescens*), the pulsating cavity represents an inflated Golgi cisterna (Schnepf and Koch 1966 a, b). When it collapses its membrane fuses with the plasmalemma and its watery contents are expelled by exocytosis. As a matter of fact the vacuole is not contractile because it disappears completely and is replaced by a subsequent cisterna. The process is the reverse of the active water intake by pinocytosis (endocytosis).

In *Vacuolaria* every 75 seconds an 8 μm large vacuole is extruded; within 30 minutes an amount of water corresponding to the full size of the cell (ϕ 26 μm) is expelled and in an hour as much as six times the cell surface of vacuolar Golgi membranes are added to the plasmalemma (Schnepf and Koch 1966 b). This means, of course, that an equivalent amount of plasmalemma must disappear at the same time. Probably it is converted into ground-plasm in a similar way as the membranes of endocytotic vesicles (Fig. 4 *b*). In view of the rapid turnover of these membranes the question must be raised whether amicroscopic elements of the plasmalemma merged in the ground-plasm can be re-used for the maturation of the membranes around the highly active Golgi cisternae.

2.3.2. Accumulation of Carbohydrates

As already pointed out, the Golgi cisternae accumulate carbohydrates. Radioactive D-glucose appears in the Golgi cisternae of root tips before the cell wall is labeled (Pickett-Heaps 1967 a, Ledbetter and Porter 1970). This accumulation must follow the same active resorption scheme as that performed by the plasmalemma. In both cases ATP is the necessary energy source.

The accumulated sugars are soluble oligomeres of the hexoses glucose, mannose or galactose and of the pentoses xylose or arabinose. A special feature are uronides such as glucuronic, galacturonic or hyalouronic acids, the polymers of which furnish highly hydrated gels with an enormous swelling capacity. Such uronides are components of hemicelluloses, mucilages and slimes. If the carboxylic groups of the galacturonides are partly methylated, they represent building units for pectic material.

In the pancreatic exocrine cells the mucin part of the secreted mucoprotein is generated in Golgi cisternae. The protein part is synthesized in the endoplasmic reticulum and, as shown by radiography, transferred from there to the condensing Golgi vesicles (Jamieson and Palade 1967).

2.3.3. Secretion of the Cell Wall Matrix

2.3.3.1. *Cell Plate*

In plant cells every cell division is accompanied by the formation of a new cell wall which functions as a diaphragm perpendicular to the mitotic spindle between the two telophasic nuclei. It appears as a floating septum in the equatorial plane of the spindle and is called *cell plate*.

The cell plate originates through the fusion of droplets visible in the light microscope. It grows centrifugally until the longitudinal walls of the mother cell are reached. In the electron microscope it can be observed that these "droplets" have ultrastructural precursors in the form of submicroscopic Golgi vesicles with less than 100 nm diameter (WHALEY and MOLLENHAUER 1963, FREY-WYSSLING et al. 1964). They coalesce in the equatorial plane and form the "droplets" just mentioned. Between the vesicles strands of the endoplasmic reticulum are visible (Plate V, Fig. N_2). These strands will guarantee direct contact of the two daughter cells by forming a system of *plasmodesmata* through the cell wall in formation. When the rim of the cell plate has reached the longitudinal walls, the two cells are separated by the so-called *middle lamella* of the future cell wall. From this moment the wall increases its thickness through the incorporation of additional vesicles on both sides (Plate V, Fig. N_4 and N_8). As a consequence of this type of growth, the surface of the young wall is not smooth but bumpy (Fig. 8).

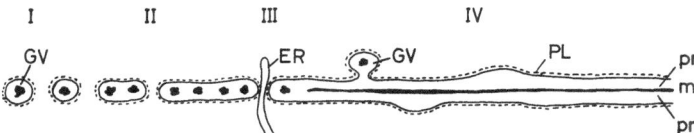

Fig. 8. Diagrammatic interpretation of the sequence leading to a new cell wall after mitosis. I. Golgi vesicles (*GV*) with a Golgi membrane and an electron dense core align and II. fuse in the equatorial plane of the cell. III. ER strand secures future plasmodesma. IV. Established wall grows in thickness by lateral fusion with additional GV. Electron dense material furnishes the middle lamella (*m*) separating the two primary walls (*pr*). The Golgi membrane of the vesicles has formed the new plasmalemma (*PL*).

By such apposition the middle lamella becomes coated on both sides by an additional wall layer which is termed *primary wall*. As each of the daughter cells produces its own thickening layer the young wall appears as a double layer of primary wall sandwiching the middle lamella. The whole trilamellate structure is the product of fused Golgi vesicles. It consists of an isotropic gel of hemicelluloses and pectic material. This amorphous and plastic mass of highly hydrated carbohydrates represents the *matrix* of the cell wall.

The center of the vesicles which furnish the middle lamella stains more deeply with $KMnO_4$ than their periphery. Probably it holds a higher concentration of uronides or pectic acid. These blackened centers can be used as a tracer for their location in the fused middle lamella. As seen in Plate V, Fig. N_7, they appear aligned in the center of the system indicating that the middle lamella is richer in highly hydratable uronic compounds than the primary walls.

Only when the primary wall is completed its plastic matrix becomes reinforced by cellulosic elementary fibrils which give the wall elasticity, mechanical strength and optic anisotropy. In this way the young wall slowly becomes birefringent so that the appearance of cellulose can be followed in the polarizing microscope. As has been pointed out (see 1.4.3.), cellulose synthesis is catalyzed by the plasmalemma. Therefore, only the plastic cell wall matrix is a product of the Golgi apparatus while the form giving principle of the wall reinforcement derives from the plasmalemma.

The most conspicuous feature of cell plate formation by Golgi activity is the transformation of Golgi membranes into plasmalemma. As seen in Fig. 8, the membrane of the Golgi vesicles coats the fusing portions of the cell plate. Later on this membrane must assume the function of the plasmalemma. As a matter of fact the plasmalemma against the new cell wall is nothing other than a coalesced sheet of membranes of Golgi vesicles.

2.3.3.2. Cell Wall Growth

When meristematic plant cells differentiate into working cells of functional tissues (palisade cells in leaves, epidermal cells, fibres, primary sieve and xylem elements), a spectacular extension of the primary cell walls is observed. As during this growth, the width of the wall does not diminish as in an extending plastic sheet, nor increase as in apposition growth of the secondary wall, the elongation is accomplished by so-called *intussusception growth.* How material is intercalated between existing structures was enigmatic until electron microscopy solved this classical problem.

It is observed that the extension growth of the cell wall is accomplished in the same way as the formation of a new wall in the cell plate. Golgi vesicles secrete their contents into the periplasmic space and their membranes fuse with the plasmalemma.

In differentiating cambial cells the Golgi vesicles may increase in size on their way to the cell surface and incorporate cytoplasmic material by invaginations. In this way multivesicular bodies coated by a Golgi membrane develop. Their contents are extruded as in the case of simple Golgi vesicles by exocytosis (Robards 1968, Robards and Kidwai 1969).

In this way two aims are achieved in one operation: the matrix for the extending wall is furnished and a concomitant surface increase of the plasmalemma is ensured. Both phenomena are local intercalation processes which show what is meant by intussusception growth. As the transfer of material is localized at the distinct spots of exocytosis, this increase in area is a mosaic growth. It must be a rapid process since cells in the hypanthium of *Oenothera acaulis* or in the epidermis of wheat roots double their surface in about 8 hours, whilst in the epidermis of the *Avena* coleoptile this time is only 30 minutes or in the filament of the grass *Anthoxanthum odoratum* even as little as 1 minute (Frey-Wyssling 1948).

The locality of intussusception growth is especially conspicuous in elongating cells which do not divide and grow only at their tip. Such apical growth performed by Golgi activity occurs in fungal hyphae (Girbardt 1969, Grove *et al.* 1970), rhizoids, root hairs (Sievers 1963), seed hairs (in cotton the cell length may reach 4 cm) and pollen tubes (the final extension depends on the length of the style, i.e. from 5 mm in grasses up to 5 cm in lilies, which means a 100 to 1000-fold growth in surface of the plasmalemma). Such cell elongation is characterized by a concentration of the process described as exocytosis in the distal pole area of the cell. As a consequence the wall of the cell tip is weaker than that of the cell mantle so that it breaks easily at that spot (FREY-WYSSLING 1957, there Fig. 21).

In pollen tubes of *Lilium longiflorum* the Golgi vesicles are so numerous that they can be isolated and purified; they contain galacturonic acid and sugars of the hot water soluble cell wall component (VAN DER WOUDE et al. 1971). In growing rhizoids of Chara the leading of the Golgi vesicles to the cell tip seems to be prescribed by the position of starch statoliths (SIEVERS 1967).

Summarizing we state that the matrix material for new cell walls (cell plate) and for their surface growth (extension growth and apical growth) is furnished by Golgi activity.

2.3.4. Secretion of Slimes and Exoenzymes

The secretion of plant slimes and mucilages proceeds in the same manner as the formation of the cell wall matrix. The only difference is that the Golgi vesicles for this purpose hold more uronides than those which yield the cell wall matrix. As a consequence such slime vesicles stain deeper with contrast producing heavy metals in electron microscopic preparations. Thus the vesicles furnishing the slime which coats the root hairs, discovered by SIEVERS (1963) in *Zea mays*, appear with a large black center or may even be completely black.

Slime extrusion is often combined with the secretion of exoenzymes. The case of the exocrinal production of mucoprotein granules in the pancreas has already been mentioned (see 2.3.2.). The slimes of insectivorous plants (SCHNEPF 1963), such as *Drosera* and *Pinguicula*, with which insects are captured display considerable proteolytic activity. As the proteases involved are synthesized in ER cisternae, the same problem as for pancreatic exocrine granules arises.

If exocytosis of large sized Golgi vesicles, visible in the light microscope, is periodical, "pulsating" vacuoles occur. Their function is water regulation and excretion.

2.3.5. Solid Secretions

The reinforcing elementary fibrils of cellulose are produced outside the metabolizing protoplast by extracellular activity of the plasmalemma (see 2.3.3.1.). However, a most striking exception to this behavior has been discovered by BROWN (1969). In the marine chrysophycean alga *Pleurochrysis* not only dissolved components but also solid fragments of the cell wall are synthesized inside Golgi cisternae in the form of scales (Plate VI, Fig. O_1). These thin solid plates are extruded by exocytosis and used in formation of the cell wall which consists of a mosaic of such scales embedded in an amorphous matrix. The scales consist of cellulose; their ultrastructure shows two laminae of cellulosic elementary fibrils, the bottom layer in concentrical and the top layer in radial arrangement (BROWN *et al.* 1970).

This special type of wall formation shows that the Golgi apparatus is not only responsible for the synthesis of the amorphous cell wall matrix, but is also capable of synthesizing cellulose and assuming morphogenetic tasks concerning the protecting wall which, as a rule, are functions of the plasmalemma.

Golgi activity competing with the plasmalemma in organizing surface protection is even more striking in the case of scale formation of flagellate algae. Green (*Heteromastix,* MANTON *et al.* 1965) and brown flagellates (*Sphaleromantis,* MANTON and HARRIS 1966; *Pyramimonas,* MANTON 1966) produce scales which cover the flagella or, if a wall is absent, even the whole cell surface. MANTON has shown that these scales are produced in Golgi vesicles. In *Pyramimonas* the scales are collected in a scale reservoir which seems also to be of Golgi origin. In any case the endoplasmic reticulum is not involved in scale formation.

The scales are liberated through a duct from the reservoir to the cell surface. It is problematic by which principle the extruded scales are arranged on the receiving surface. This question is the more enigmatic when two types of scales are produced as in *Sphaleromantis* (Plate VI, Fig. O_2). In this organism many small scales and a reduced number of elongated scale rods are produced. The first morphogenetic riddle is how Golgi vesicles are informed how many of each scale type must be synthesized and the second how, by extracellular activity, the two scale types are arranged in a predetermined pattern.

A similar type of secretion of solid surface building elements is observed in chrysophycean and xanthophycean algae with hairy flagella. There the "hairs" appear again in cisternae as bundles of parallel tubules which are extruded and attached to the flagellum. However, as these tubules are of a proteinaceous nature, they are not produced in Golgi vesicles but in the perinuclear space by the system of the protein synthesizing endoplasmic reticulum (LEEDALE *et al.* 1970). Membrane bounded cisternae detach from the nuclear envelope and migrate to the cell surface for exocytosis (see 4.2.2. and Plate IX, Fig. W, p. 83).

2.3.6. Translocation

NORTHCOTE (1971) is of the opinion that the main function of the Golgi system is translocation. This interpretation gives priority to the distribution of Golgi vesicles over synthesis of physiologically important compounds. Of

---

Plate VI. Fig. O. Solid products of Golgi activity; O_1 formation of cellulosic scales in Golgi cisternae (arrows) of *Pleurochrysis,* their extrusion by exocytosis and incorporation into the laminated cell wall, $\times 68,000$ [courtesy of R. M. BROWN, JR., from J. Cell Biol. **41**, 109 (1969)]. O_2 formation of scales and rods (double arrow) in cisternae evolved from Golgi vesicles (simple arrow) of *Sphaleromantis,* $\times 20,000$ [courtesy of I. MANTON and K. HARRIS, from J. LINNEAN, Soc. London **59**, 402 (1966)].

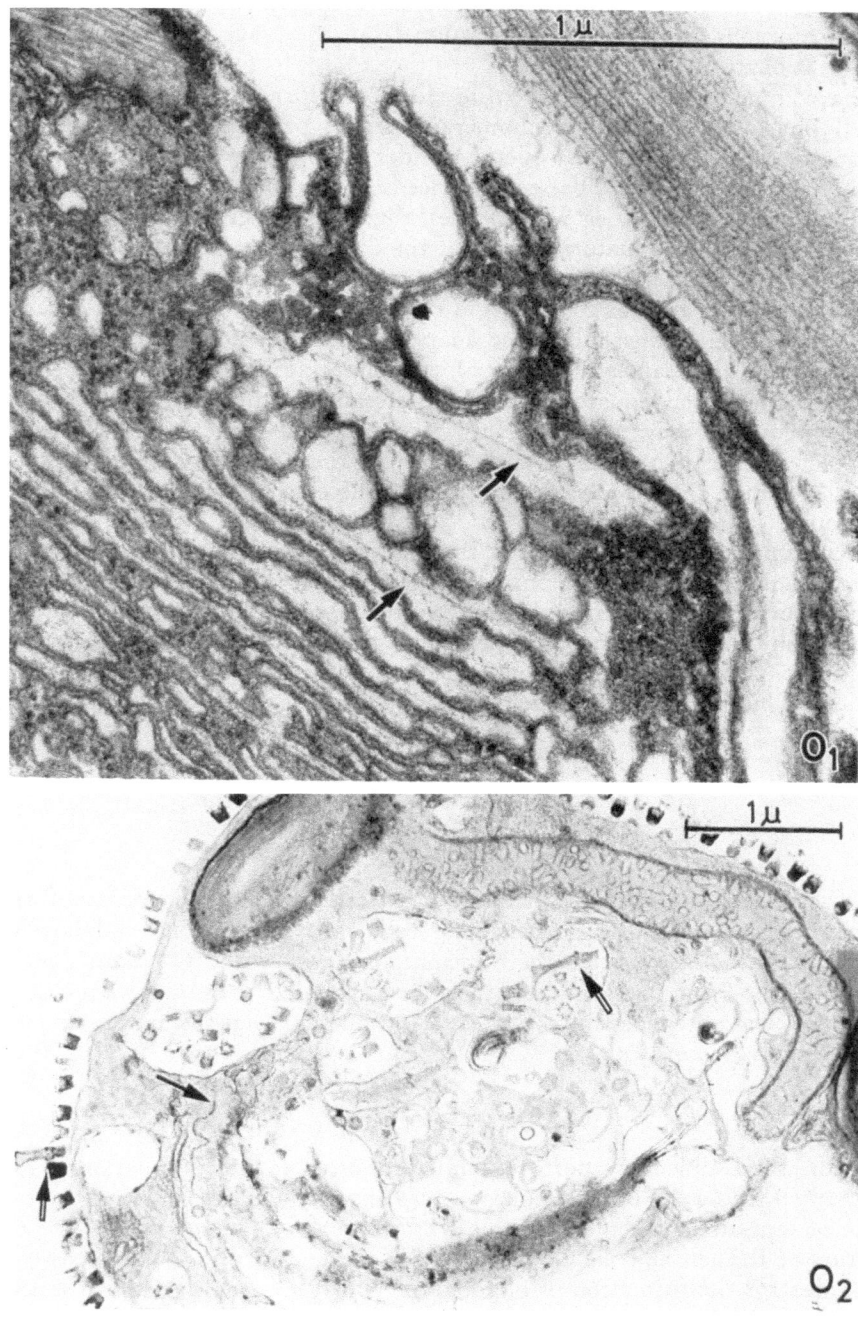

Plate VI.

course, the function of cellular secretion involves both activities: production of the secretions and their transportation from the place of synthesis to the place of exocytosis.

The combination of these two functions has opened quite a new avenue in cell physiology, because a number of urgent problems arise. Especially the question how the Golgi vesicles wander to the sites of the cell surface where their contents must be secreted needs a solution (see 1.4.2.). Another thing we want to know is what forces align the Golgi vesicles which fuse during mitosis in the equatorial plane of the dividing cell. LÓPEZ-SÁEZ (1964) considered a passive mechanism for the movement of the vesicles: When the mitotic apparatus extends towards the two poles, a piston pressure is exerted on the cytoplasm against which the anaphase chromosomes and the telophase nuclei move. This causes a stream of cytoplasm along the the walls in the direction of the cell equator, where the streams from the two poles meet. As a consequence they must turn into the equatorial plane where they align the Golgi vesicles met with on their way. As the supposed stream comes to its end in that plane the vesicles would stay there as observed in the electron microscope (Plate V, Fig. N_2).

As ingenious as such mechanistic theories may be, they will never be able to explain the essence of cell physiology which consists in *active* movements and the pursuit of an aim. As such activities need an energy source and an information program, physical phenomena such as concentration, pressure or electric gradients cannot be considered as the cause of physiological reactions which are rather the effect than the cause of metabolic molecular processes.

2.4. Identity with the Plasmalemma

2.4.1. Homology

It has been shown that in mitosis of plant cells the Golgi membrane forms the new plasmalemma along the young cell wall (see 2.3.3.1.) and how on exocytosis of Golgi vesicles their membrane fuses with the plasmalemma (see 2.3.3.2.). From these facts it must be concluded that the two types of membranes are ontogenetically related, i.e. that the Golgi membrane and the plasmalemma are *homologous*.

2.4.2. Analogy

At the same time they perform similar or even identical functions. This is shown in Table 2. Both membranes are involved in resorption by active intake of ions and molecules. In this way the plasmalemma provides for the nutrition of the cell and the Golgi cisternae for the resorption of sugars which are needed for the formation of oligopolymers of the carbohydrates in mucin, slimes, hemicelluloses and/or pectic material. Another common activity concerns elimination processes which are performed by exocytosis whereby the vesicular membrane fuses with the plasmalemma. Of course, resorption (e.g. of polypeptides by ER membranes) and exocytosis are also characteristic for other biomembranes, but not in such close cooperation with the plasmalemma as stated in Table 2.

The most striking similarity between Golgi and cell membranes is, however, the common function of both types in providing solid carbohydrate components for the surface protection of the cell: The plasmalemma produces elementary fibrils for the reinforcement of the cell wall and the Golgi apparatus synthesizes solid scales (Plate VI, Fig. O_2) or in the case of *Pleurochrysis*, even complete cellulosic components of the cell wall (Plate VI, Fig. O_1).

Table 2. *Analogous Functions of Plasmalemma and Golgi Membrane in Plants*

Functions	Plasmalemma	Golgi membrane
Resorption	by uphill intake of inorganic ions and micromolecules by endocytosis of macromolecules (*e.g.,* ferritin)	by uphill intake of micromolecules (sugars) and macromolecules (proteins)
Elimination	by exocytosis	by exocytosis
Surface protection	synthesis of wall fibrils (cellulose, glucan, chitin) formation of myelin sheath around nerves	synthesis of wall matrix synthesis of cellulose in *Pleurochrysis* formation of scales around flagella

This review shows that plasmalemma and Golgi membranes have the same functions! The only difference is that the activity occurs in one case at the cell surface and in the other inside the cell. In comparative anatomy, organs which have the same functions fall within the concept of analogy. Therefore, Golgi apparatus and plasmalemma are *analogous* organelles.

2.4.3. Identity

As a consequence, we state that the two types of membranes under discussion are not only ontogenetically but also functionally alike; they are at the same time homologous *and* analogous which means that they are *identical*.

This is an important conclusion, because it demonstrates that the Golgi apparatus must be considered not only as an organelle for elimination processes (secretion, excretion) but also as an organelle for plasmalemma production. The membranes of mature Golgi cisternae and of budded off Golgi vesicles must be visualized as internal plasmalemma which represents an enormous surface increase of that important organelle.

This approach is especially obvious in the case of the spectacular cell extension in certain tissues of higher plants. As explained under 2.3.3.2., the epidermal cell of the oat coleoptile may increase its surface almost 40 times and a cotton hair even 70 times a day (FREY-WYSSLING 1948). For this growth the Golgi apparatus not only furnishes the considerable amount of matrix for the elongating wall, but at the same time the necessary material for such a rapid surface increase of the plasmalemma. In *Vacuolaria* (see 2.3.1.) Golgi activity incorporates within the even shorter time of only one hour membranes of six times the cell surface into the plasmalemma (SCHNEPF and KOCH 1966 b).

As a rule the number of dictyosomes is higher in growing than in resting cells. This may be in connection with the need of a more highly developed system of internal plasmalemma formation during the period of intensified metabolism.

The conclusion of this section is that Golgi membranes and plasmalemma are ontogenetically and functionally alike (Table 2). The Golgi apparatus seems to function as a system for cell-internal plasmalemma production. It is of paramount interest whether the identity of these organelles stated by ultrastructural morphology will also hold for their biochemistry, which question must be solved by detailed protein and lipid analysis of the isolated membranes (SITTE 1969 a, KORN 1969, WHALEY *et al.* 1971, p. 2).

3. Endoplasmic Reticulum (ER)

3.1. Ultrastructure

In 1945 PORTER *et al.* discovered a fine network in the endoplasm of chicken fibroblasts which was called *endoplasmic reticulum* and abbreviated ER (PORTER 1948). In thin sections it appears as profiles of isolated strands, elongated vesicles and cisternae, but serial sections, electron stereographs and freeze etchings (FREY-WYSSLING and MÜHLETHALER 1965, there Fig. *E–20*) show that these elements intercommunicate spatially and form a real reticulum.

The dimensions of the larger cisternae are such that they ought to be visible in the light microscope and in special instances this is actually the case when using phase contrast (FAWCETT and SUSUMO 1958). Nevertheless the endoplasmic reticulum is considered as an ultrastructural organelle because its details are only resolvable in the electron microscope.

3.1.1. ER Membrane

The elements mentioned are coated by a unit membrane which is considerably thinner than the plasmamembrane and mature Golgi membranes. In homogenized cells the elements round up and form spheres with a diameter around 0.15 µm which were termed *microsomes* by CLAUDE (1946). The microsomal membranes can be analyzed and are found to hold two thirds protein and one third lipids (Table 1, p. 23).

The ER elements contain a clear electron transparent fluid with almost the same refractive index as the groundplasm. This is the reason why the reticulum remained hidden to classical cytologists until the phase microscope was available. The fluid contains salts and soluble proteins. Therefore, it is comparable to serum, and corresponds, together with the sap in Golgi cisternae and in the thylakoids of chloroplasts, to the *enchylema* of classical cytology.

The surface of the ER membrane is either smooth or rough if it is covered with electron-dense particles which are called *ribosomes* (see 3.5.1.) because they hold up to 65% ribonucleic acid (SITTE 1969 b, p. 24, KARLSON 1970, p. 115, WHALEY *et al.* 1971, p. 13). There are regions in the cell where all profiles of the endoplasmic reticulum are smooth and others where they look rough. It is rare that the same ER cisterna is rough on one side and smooth on the other as reported of sieve elements in wheat seedlings (PICKETT-HEAPS and NORTHCOTE 1966 a). However, this case is characteristic of the nuclear envelope where only the outer membrane may be covered with ribosomes.

3.1.2. Nuclear Envelope

The difference between the primitive prokaryotes (bacteria, blue-green algae) and the eukaryotes consists in the absence or existence of a delimitation between the material carrying the genetic code and the cytoplasm. Originally it was thought that this boundary was due to a surface film or a membrane, but electron microscopy revealed a more complicated system of two membranes separated by a perinuclear space 10–20 nm wide and filled with enchylema. This double membrane is called *nuclear envelope*.

The nuclear envelope is perforated by pores of considerable dimensions (e.g. 20 nm in yeast). Depending on whether thin sections strike the nuclear surface radially or tangentially the pores are seen in profile or from the top. Survey pictures of the whole nuclear surface can be obtained by freeze etching (MOOR and MÜHLETHALER 1963). An example is shown on Plate VII, Fig. P_1. In this way it is found that the pores are not evenly distributed. Some regions of the surface may be rich in such openings while others are devoid of them (PEDERSEN 1972). They also do not seem to be permanent, because they may appear as if recently abolished. It is plausible that the regions with pores may change their position on the nuclear surface according to functional needs. As a consequence their number should be evaluated with regard to the physiological activity of the cell. Without such a physiological approach the comparison of pore numbers in the nuclear envelope of different biological material is problematic. On an average 10–30, and exceptionally up to 100 pores/μ^2 have been observed (THAIR and WARDROP 1971).

The pores guarantee intimate contact between nucleoplasm and groundplasm. However, the pores have no open bore. If it were so, the plasm inside and outside the nuclear envelope would be continuous, so that there would be no need to distinguish between nucleoplasm and groundplasm, which are both finely granular. Phylogenetically such an identity would be conceivable, because it is evident that the plasm around the genetic system of the prokaryotes was originally the same as that which became separated from it by the process of compartmentation. The two plasms have not only the same granular aspect but also the common capacity of producing microtubules for spindle fibres which in some cases, e.g. yeast may arise inside the nuclear envelope. An open pore would be welcome to explain the transfer of messenger RNA through the nuclear envelope (Fig. 9).

However, as the pores are blocked by electron-opaque material, exchanges cannot proceed by mere diffusion based on concentration gradients. Since the entrance into the nucleus, not only of ions (Ito and Loewenstein 1965), but also of colloidal gold particles coated with poly-L-proline or poly-L-lysine with a diameter up to 14 nm (Feldherr 1965) has been observed, exchanges

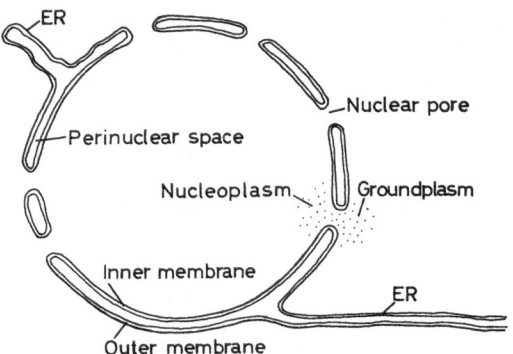

Fig. 9. Diagram of the nuclear envelope.

between groundplasm and nucleoplasm must include energy consuming metabolic processes. The complicated internal ultrastructure of the pores points in this direction.

Some authors have observed a fine diaphragm and many others describe a ringshaped structure or annulus around the openings of the pores in oocytes (Afzelius 1955) or a special pore margin in hepatic cells (Watson 1959). With the Markham reinforcement technique which enhances radial symmetry of ringshaped structures (Markham et al. 1963), it can be shown that the annulus is not a continuous rim but a circular arrangement of eight particles of about 20 nm diameter on either side of the pore (Fig. 10 a) which is on an average 72.5 nm wide (Franke and Kartenbeck 1969, Franke and

Plate VII. Fig. P. Nuclear envelope of yeast nuclei; P_1 erratic distribution of nuclear pores, $\times 24,000$ (courtesy of K Mühlethaler); P_2 cross section of freeze-etched nuclear envelope interrupted by a pore, the perinuclear space between the two ER membranes is considerably wider than in chemically fixed and dehydrated preparations, $\times 180,000$ (courtesy of H. Moor). Fig. Q. Plasmodesmata cross-sectioned, cell wall in the perianth of Narcissus flower (cp. Fig. 13), $\times 70,000$ (phot. H. Kuhn). Fig. R. Ontogeny of spherosomes; R_1 ER strands bud off prospherosomes (slightly shrunken by dehydration due to chemical fixation) in the epidermis of onion scales, $\times 18,000$ [from J. Ultrastructure Res. 8, 506 (1963)]; R_2 sphero-somes in the seed of Ricinus showing spectacular lipophanerosis of the middle stratum of the original ER unit membrane, the inner black stratum is seen within the spherosome (arrows), P prospherosomes, $\times 16,000$ [courtesy of A. M. Schwarzenbach, from Cyto-biology 4, 145 (1971)].

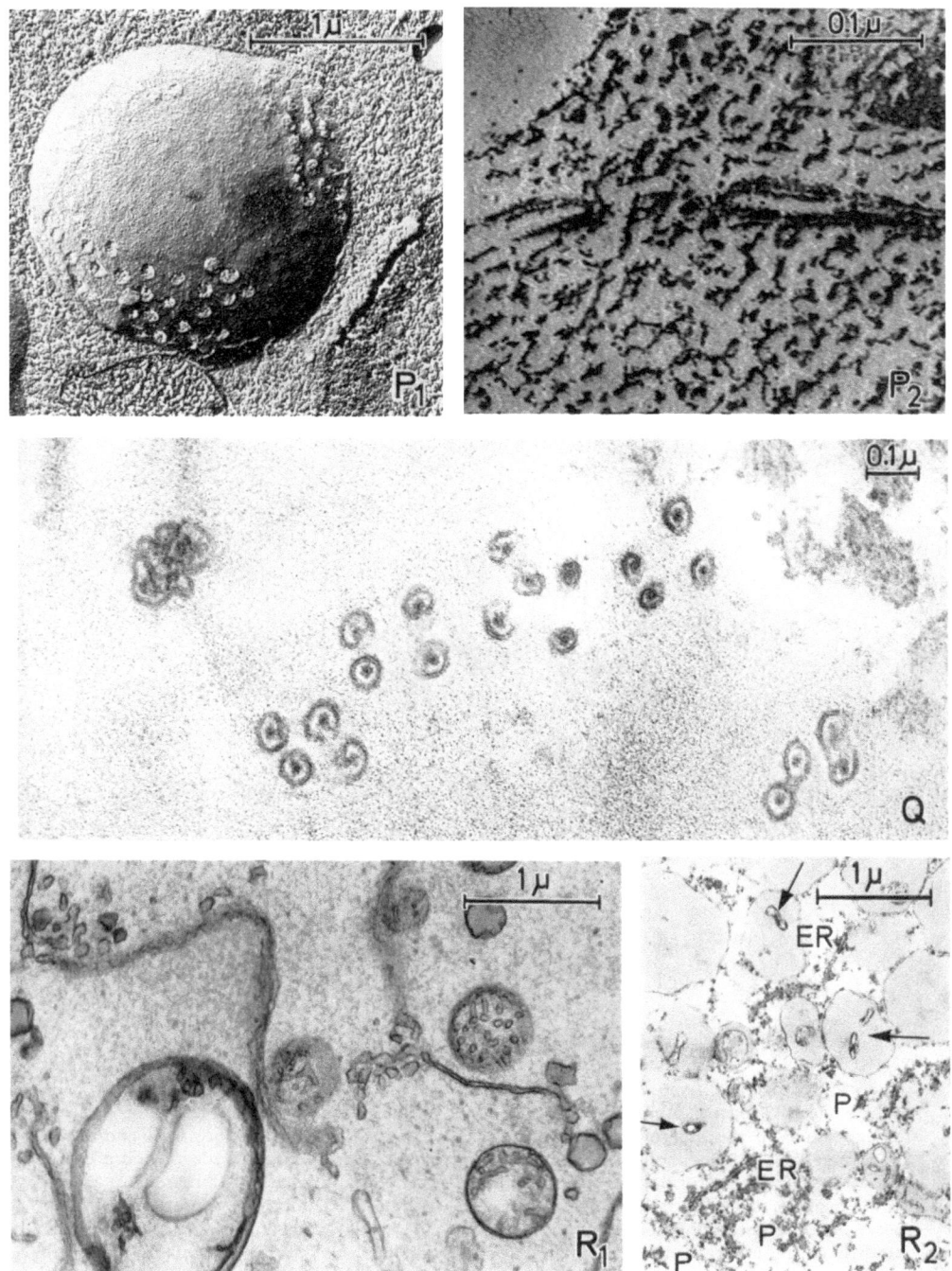

Plate VII.

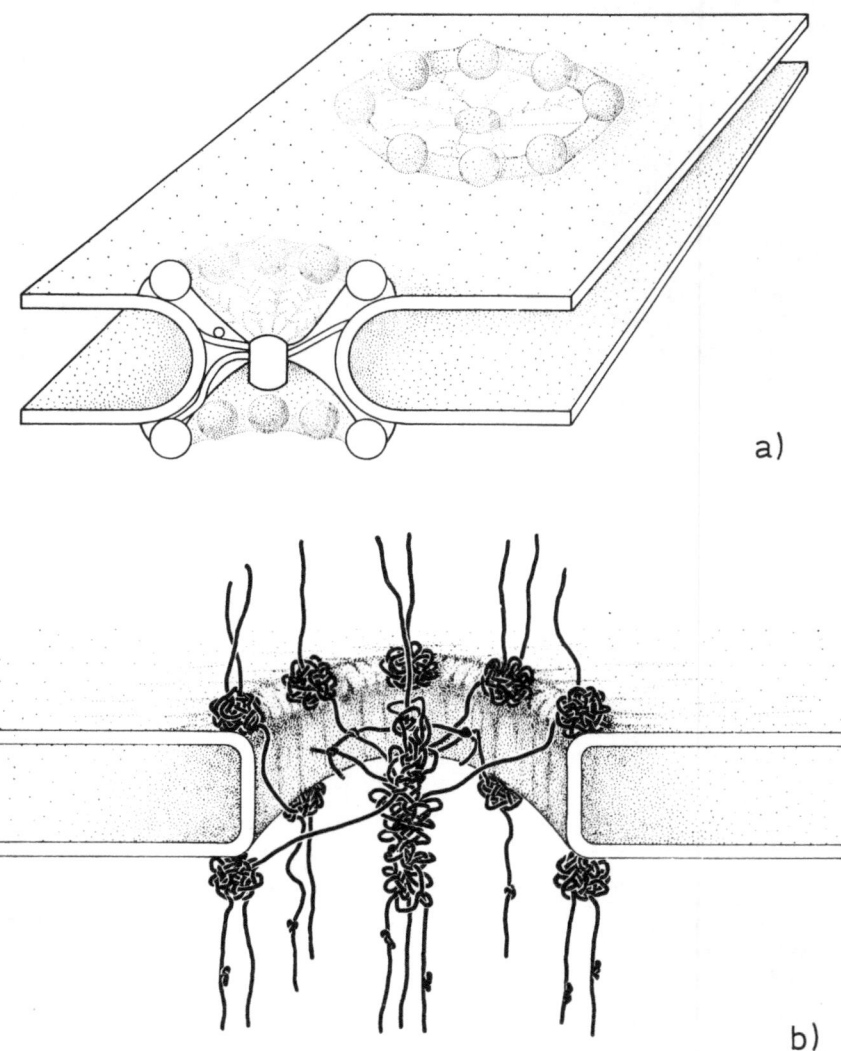

a)

b)

Fig. 10. Nuclear pores. Diagrammatic concepts of *a*) globular and *b*) fibrillar pore complex structures as derived from electron micrographs of a diversity of animal and plant cells. Eight regularly spaced annular granules lie upon either pore margin. A central granule or rod is located in the innermost part of the pore. It is attached to the pore wall and/or the annular granules. Amorphous material projects from the pore wall towards the pore center and takes part in the formation of an equatorial layer ("diaphragm"). For reasons of clarity the amorphous material is not included in the diagrams. The fibrillar aspect of Fig. 10 *b* shows the different filamentous structures observed: (1) fibrils attached to the annulus, (2) fibrils attached to the central granule, (3) fibrils protruding into the outer groundplasm, (4) fibrils making up an "inner ring", (5) filaments connecting the central granule to the pore wall and the annular particles [courtesy of W. W. Franke, from Z. Zellforschung 105, 405 (1970)].

SCHEER 1970). FRANKE (1970 a) shows that these features are universal in plant and animal cells (root tips, leaves, liver parenchyma, oocytes, HeLa cells).

The pore has the shape of an hour-glass with a central granule or *central dot*. This particle has fibrillar connections with the annular granules from which again fibrillar material radiates towards the nucleoplasm and the groundplasm (Fig. 10 *b*). As the annular and the central granules are macromolecules with coiled molecular chains, FRANKE (1970 a) visualizes not only a granular but also a fibrillar model of his findings (Figs. 10 *a, b*).

The central granule shows the absorption and staining properties of nucleic acid. Therefore, it seems to represent RNA particles leaving the

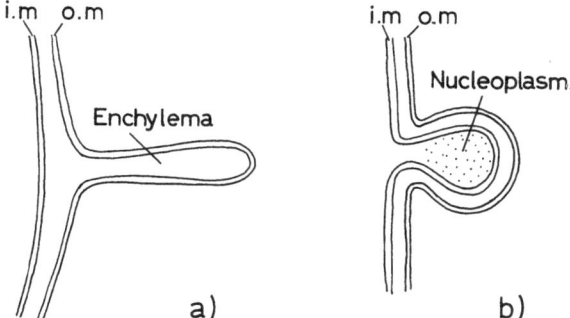

Fig. 11. Nuclear evaginations of *a*) outer membrane (*o.m.*) containing enchylema (ER strand) and *b*) both the outer (*o.m.*) and the inner membrane (*i.m.*) containing nucleoplasm (Bell bodies).

nucleus. The rate of this transport can be calculated based on the increase of cytoplasmic RNA during exponential growth, the number of pores per nucleus in the order of 10^4 to 10^7 and under the assumption of a mean molecular weight of 1.15×10^6 for RNA which corresponds to the average amount of ribonucleic acid in the 50 S and 30 S parts of the ribosomes containing 16–18 S or 25–28 S RNA respectively (Fig. 16). In this way FRANKE (1970 b) finds that 1 to 24 RNA molecules per minute may pass through the nuclear envelope.

As the nuclear pores are certainly not static openings but dynamic locks which are formed only temporarily as the circumstances may require, it would be of interest to know how many RNA molecules are transported through a pore during its lifetime or whether for the transfer of every single macromolecule the complicated poral ultrastructure must be installed. At any rate the electron micrographs available are only snapshots of an action which is subject to periodical morphological changes.

Another question is whether the inner membrane of the nuclear envelope (lining the nucleoplasm) and the outer membrane (lining the groundplasm) are of identical structure. As these two membranes meet in the canal of the nuclear pore (Fig. 10), it seems that there is but one type of continuous membrane. On the other hand they behave differently when the nuclear envelope forms evaginations. Such processes may be the result of dynamic activity of the nuclear envelope in a similar way as described for the plasmalemma (see 1.2.). However, besides such transient formations, the nuclear envelope is also involved in the production of new structural elements.

In this respect two cases must be distinguished. Either only the outer membrane evaginates or both the inner and the outer membranes are involved in the formation of protuberances (Fig. 11). In the latter case bodies with a double membrane containing nucleoplasm are formed (Fig. 11 b). They were discovered by Bell and Mühlethaler (1962) and will therefore be called Bell bodies. Their cytological status is controversial (see 3.3.).

If only the outer membrane evaginates, ER strands are formed (Fig. 11 a). This means that the reticulum is continuous with the nuclear envelope and, as a consequence, the perinuclear space must contain ER enchylema. Therefore, the nuclear envelope is a part of the ER system. However, only its outer membrane may carry ribosomes whilst the inner membrane is always smooth.

3.2. Ontogeny

The fact that the nuclear envelope can generate ER elements raises the question whether the whole ER system of the cytoplasm derives from the nuclear envelope or vice versa. As the nuclear envelope "disappears" during mitosis, its ontogeny can be studied when it is reformed in the telophase. Such investigations show that there is no de novo formation of nuclear membranes but a fusion of ER vesicles into which the nuclear envelope disintegrated during the prophase (Porter and Machado 1960). Other vesicular elements which cannot be distinguished from those of the disintegrated nuclear envelope regenerate the ER system in the cytoplasm of the two daughter cells. In this way ER membranes persist during the series of cell generations so that a de novo formation of this type of membrane by the groundplasm seems unlikely.

The question of the relation between the ER system and the nuclear envelope can also be discussed from a phylogenetic point of view. Since the prokaryotes have no nuclear envelope around their genetic material, the presence of an endoplasmic reticulum would characterize this organelle as the precursor of the nuclear sheath. However, no typical ER cisternae are described in prokaryotes so that these two membrane systems probably appeared simultaneously in cell phylogeny.

It is not known why the compartmentation which separates and differentiates the nucleoplasm from the cytoplasm became necessary. The view that the important genetic code, as a most sensitive constituent, had to be protected from the metabolic and ontogenetic manifestations in the cytoplasm of differentiating cells is dubious, because it is just the very task of the nucleus to organize and control these events for which purpose the nuclear envelope is perforated by its dynamic system of pores. The function of the nuclear envelope might, therefore, be less to fulfil the need of protection than to provide the nucleus with the necessary metabolites furnished by the enchylema. In this case the nuclear envelope would have a nutritional function.

This idea can be supported by the fact that no envelope is necessary during the stagnancy of DNA synthesis in the anaphase and telophase, and by a cytological speciality in certain algae (MCBRIDE and COLE 1969, URBAN 1969, PICKETT-HEAPS 1970) where a ring of ER profiles is found around the nuclear envelope (Fig. 12). This additional shell seems to sustain the function of the nuclear envelope which may have arisen phylogenetically in a similar way as this "outer envelope" when the uncoated nuclear region of the prokaryotes became surrounded by ER elements.

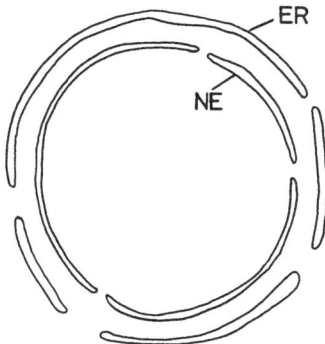

Fig. 12. Perinuclear layer of ER cisternae in red algae. *ER* endoplasmic reticulum, *NE* nuclear envelope.

3.3. Problematic Homologies

The two membranes around plastids and mitochondria are different. On ontogenetic grounds from prokaryotes the inner membrane of the plastidal envelope can be considered as homologous with the plasmalemma (see 1.5.2.) and, if there is any phylogenetic relationship between plastids and mitochondria this would also hold good for the inner membrane of the mitochondrial envelope. For their outer membrane, however, no homology can be derived from ontogeny. If during phylogenesis of the plastids the region of the thylakoids in prokaryotes (i.e. the chromoplasm of classical cytology) had been coated in a similar way as the nuclear region, however, not by a double but by a simple film, the outer membrane of this organelle may have conserved its ER nature. The fusion of ER membranes with the outer plastidal membrane which has been observed in the archegonium of *Ginkgo biloba* (CECCHI FIORDI and MAUGINI 1972) points in this direction.

These considerations are essential for a discussion concerning the formation of Bell bodies on the nuclear surface in the egg cell of ferns. Before fertilization the mitochondria and plastids of the egg disintegrate and seem to be replaced by globular bodies which originate through evagination of the nuclear envelope (see 3.1.2.), whereby both the outer and the inner membranes are involved (Fig. 11 *b*). The bodies given off by budding contain nuclear material consisting of nucleoplasm, nucleolar substances (Bell 1970) and DNA (Bell and Mühlethaler 1964). As this genetic material is nuclear DNA which differs from plastidal and mitochondrial DNA, it is uncertain whether the Bell bodies can be initials for chloroplasts or mitochondria. Diers (1965) tried to prove by serial sections that the pictures published did not represent bodies separated from the nucleus but only profiles of convolute nuclear processes. Nevertheless, many of these protuberances swell and become independent whereupon the inner membrane seemingly produces plastidal thylakoids (Bell et al. 1966).

This would only be possible if the inner membrane of such initials and, therefore, also the inner membrane of the nuclear envelope were homologous with the plasmalemma. However, both the outer and the inner membranes of the nuclear envelope derive from the ER membrane as demonstrated through the reconstruction of the envelope after mitosis. For an organellographic understanding of the findings of Bell, a transformation of the faculties of the ER membrane into those of the plasmalemma ought to be postulated in the case of the inner membrane of the nuclear envelope. This calls for a clarification of the interrelations between ER and PL membranes.

3.4. ER and PL Membranes

Many authors claim that membranes of the endoplasmic reticulum (ER) may fuse with the plasmalemma (PL). This would mean that the ER cisternae are in open communication with the surrounding medium which, in this way, would have free access even to the perinuclear space. A well known model of this view is the illustration by Robertson (1959) which has been reproduced in many textbooks. This model is based on the unit membrane concept which claimed a uniform structure for all cytomembranes and, apart from individual vesicles, their uninterrupted continuity. Even the outer membrane of the mitochondria, the homology of which is uncertain, derives according to this concept from the plasmalemma.

Although this simplification, convincing as it was in those days, had to be generally abandoned, there are still cytologists who believe in a possible fusion of the plasmalemma with, or a formation of plasmalemma by, the ER system (Hepler and Jackson 1968). However, in our experience with cells of higher plants, we have never met with such a confluence of ER membranes with the plasmalemma. It is true that ER strands can run very closely along the plasmalemma (Fig. 17 c); but in clear electron micrographs no real contact of the membranes is ever visible. Pictures which were meant to show the continuity of the nuclear envelope with the plasmalemma by means of an ER diverticulum in the fungus *Stilbum zacalloxanthum* or in the archegonium of *Ginkgo biloba* (Moore and McAlear 1961, there Fig. 3; Maugini and Cecchi Fiordi 1971, there Fig. 6) are blurred at the very place where ER and PL membranes should meet and make contact.

The best proof for the incompatibility of these two types of membrane are the *plasmodesmata* which join adjacent plant cells through their separating primary wall (Frey-Wyssling and Mühlethaler 1965, Fig. E–8).

A section perpendicular to such a plasmic strand (60 nm wide) shows two concentric cytomembranes. The outer (10 nm wide) is the plasmalemma while the inner one (7.5 nm wide) is the ER membrane of the ER channel (Fig. 13) which provides continuity of the ER system on either side of the wall. Its bore, visible in Plate VII, Fig. Q, as a black dot, measures 5 nm in diameter. During differentiation an exchange of enchylema between adjacent cells through these capillaries is possible. Yet, in full-grown cells they may

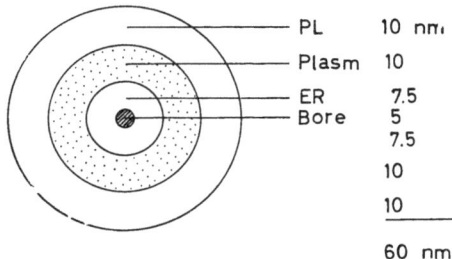

Fig. 13. Cross section of plasmodesma (cp. Plate VII, Fig. Q).

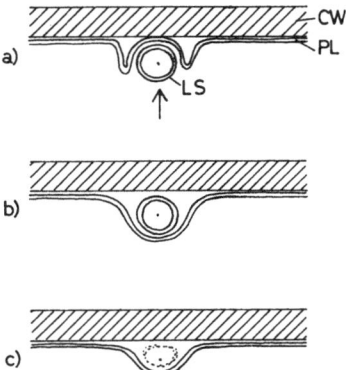

Fig. 14. Extrusion of a lysosome (*LS*) in *Neurospora* (according to MATILE *et al.* 1965). *a*) The ER membrane of the lysosome does not fuse with the plasmalemma (*PL*). *b*) Lysosome in the pericellular space between plasmalemma (*PL*) and cell wall (*CW*). *c*) Breakdown of the lysosomal membrane (cp. Fig. 19).

appear blocked, especially when the cells develop a considerable secondary wall. Nevertheless, the plasmic bridge persists as a layer of groundplasm between the PL and the ER membrane. The plasmic layer is only 10 nm wide and this small distance is held over the whole length of about 1000 nm. Never does any fusion of the two membranes occur.

Another example of incompatibility is the extrusion of ER bounded lysosomal vesicles through the plasmalemma (Fig. 19). In contrast to exocytosis of Golgi vesicles, the ER vesicles keep their membrane which does not fuse with the plasmalemma during the secretion process (Fig. 14) and disappears only after having left the protoplast (MATILE *et al.* 1965).

In the same way, when Golgi vesicles are incorporated into the ER derived vacuole, no fusion of the PL-homologous Golgi membrane with the ER-homologous membrane of the tonoplast occurs (Matile and Moor 1968).

Notwithstanding this apparent inability of joining, PL and ER membranes are interrelated. As pointed out under 2.2., Golgi cisternae can originate by fusion of vesicular ER elements (Plate V, Fig. *M*). However, the ER membrane undergoes a considerable transformation in the dictyosome before it can function as a Golgi membrane actively producing secretion vesicles. This transformation or maturation process becomes manifest by a striking thickening of the membrane when the cisterna shifts from the proximal to the distal pole of the dictyosome (Fig. 7).

In yeast the Golgi apparatus is an ephemeral structure which in most cells cannot be found. It is produced by proliferating ER elements. Similar vesicles of an ER field, at the place where a new bud will sprout, are involved in the formation of the new cell wall (Moor 1967 a, Matile et al. 1969, 1971).

In the case of ascomycetes several authors have described how, in the ascus of these fungi, the plasmalemma of the spores derives from ER elements. The ascal cytoplasm which contains 4 or 8 haploid nuclei is comparted into future ascospores by flat ER cisternae (Fig. 15 a) which fuse in a similar way as in the reconstitution of the nuclear envelope after mitosis. The result is that the protoplast of the future ascospore is delimitated by a double membrane called *sacculus* (Carroll 1967), the inner membrane of which assumes the function of the future plasmalemma of the spore.

However, the ER membrane seems to undergo a transformation. In *Pustularia* the element which accomplishes the separation of the sporal from the ascal cytoplasm originates opposite to a special process of the nucleus (Fig. 15 b). It assumes the shape of a Golgi cisterna with an inflated rim called *ampulla* (Schrantz 1966) which extends by peripheral growth and encircles the nucleus. Later its inner membrane becomes the plasmalemma of the spore and the contents of the space between the inner and the outer membranes are densified so forming the sporal cell wall. The material in the extending cisternal space is thus produced by ER elements (Schrantz 1967) in a similar way as in yeast (Fig. 15 c).

Also the wall of the hat-like ascospores of *Hansenula* originates from a hat-shaped ER cisterna the space of which contains mucopolysaccharides that become the cell wall and the inner cisternal membrane acts as plasmalemma (Black and Gorman 1971).

It seems that in fungi there is no (or a not as yet) visible differentiation of ER and Golgi elements. In *Gilbertella* rings of ER cisternae resemble a Golgi apparatus (Bracker 1967) and, as mentioned, in yeast a field of ER vesicles functions in the manner of Golgi vesicles for local growth of the cell wall (Moor 1967 a). These vesicle not only secrete matrix material (mannan) through the plasmalemma but also a lysosomal hydrolase (β-1,3-glucanase) which is necessary for the local softening of the cell wall where the yeast cell generates a bud (Matile et al. 1971, Cortat 1971).

In higher plants these different functions have been divided between more

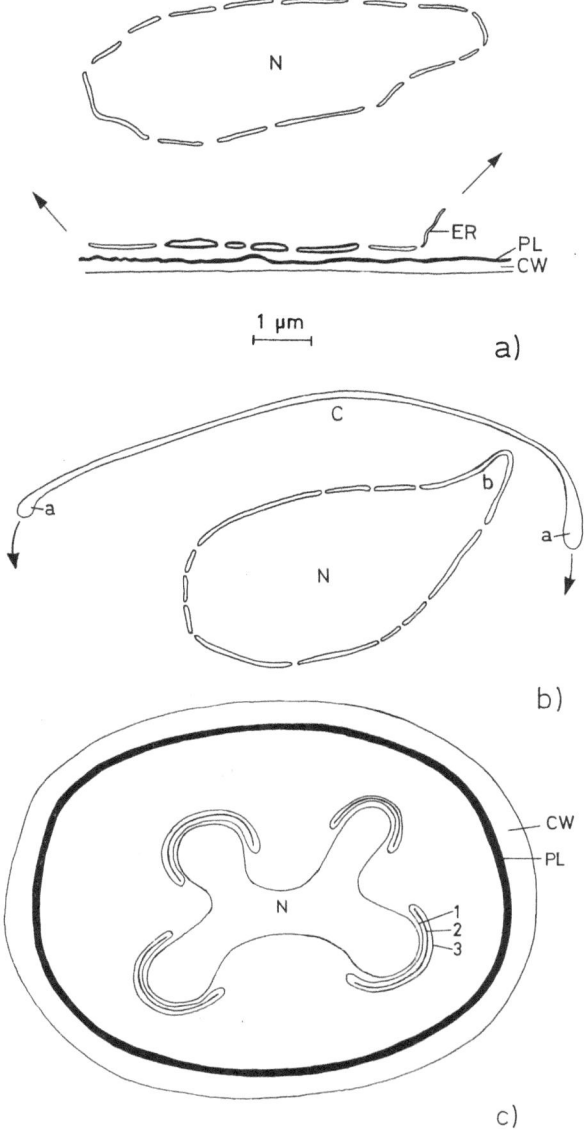

Fig. 15. Ontogeny of the plasmalemma in asco-spores. *a*) In *Saccolobus* produced by fusion of flat ER cisternae, the membrane of which grows in width. The nucleus (*N*) is encircled (arrows) by this ER double membrane the inner layer of which differentiates into the plasmalemma of the asco-spore. No invagination of the Plasmalemma (*PL*) under the cell wall (*CW*) of the ascus is observed (drawn from CARROLL 1967, Fig. 7). *b*) In *Pustularia* produced by an ER cisterna (*C*) which has the shape of a Golgi saucer (*a* = ampulla) growing around the haploid nucleus (arrows). The cisterna arises opposite a special nuclear "horn" (*b*). The contents of the cisterna are transformed into the cell wall of the asco-spore (drawn from SCHRANTZ 1966, Fig. 1). *c*) In *Saccharomyces* (yeast). Four coherent haploid nuclei. *CW* cell wall, *PL* plasmalemma of ascus.

(1) inner membrane of cisterna → PL of asco-spore,
(2) contents
(3) outer membrane } of cisterna → CW of asco-spore.

specialized organelles which differ not only functionally but even present a morphologically different particle pattern in their membranes (Matile and Moor 1968).

It could be that in higher plants, the structural differences of the PL and ER membranes are genetically fixed to such a degree that the shift ER → PL cisternae observed in lower plants is no longer possible. In this case ER and PL membranes could well be interrelated phylogenetically but not ontogenetically. Whether there is a real lack of ontogenetic homology in higher plants can only be established by further research. On the other hand the phylogenetic homology demonstrated

$$ER\ membrane \rightarrow Golgi\ membrane \rightarrow PL\ membrane$$

supports the view that the Golgi system is indeed an apparatus for plasmalemma formation (see 2.4.3.). Not only does it produce vesicles for the expansion in area of the existing plasmalemma but it is also capable of transforming ER into PL membranes.

There are not only ontogenetic but also chemical differences between ER and PL membranes. The enzyme complexes they carry are not the same and their specific proteins can be tested with the methods of immunology (Gitzelmann et al. 1970). The plasmalemma is destroyed by basic proteins which is not the case for the vacuolar membrane. The vacuoles of naked yeast protoplasts can be freed in this way because their tonoplast which is ontogenetically an ER membrane (see 3.6.) resists such a treatment (Wiemken 1971).

The negative electric charge on the surface of the PL membrane is considerably denser than that of the ER membranes. As a consequence PL fragments and ER microsomes of a homogenized suspension can easily be separated from each other in an adequate electrophoresis chamber (Heidrich 1972). However, no clear electrophoretic separation seems possible between ER and Golgi membranes. This uncertainty may be related to the fact that a dictyosomal stack presents a series of membrane transformations, in that the proximal cisternae pass through a process of maturation before they acquire PL properties and PL homology at the distal pole.

3.5. Functions

3.5.1. Synthesis of Proteins

The main task of the rough endoplasmic reticulum is protein synthesis. Whereas the Golgi apparatus is capable of producing oligomeric carbohydrates from simple sugars, the ER system polymerizes amino acids into specific polypeptides. For the production of proteins associated with carbohydrates, as for instance the synthesis of mucoprotein in the pancreas, the two organelle complexes cooperate (Jamieson and Palade 1967).

The enzymes for the synthesis of polypeptidic chains are located in *ribosomes*. These osmiophilic granules are found either in the groundplasm or attached to the strands of the endoplasmic reticulum. Their high content of ribonucleic acid (see 3.1.1.) gives them basiphilic properties so that their accumulation in the plasm can be disclosed in the light microscope by means of basic dyes. Such basiphilic regions of the cell are known as *ergastoplasm.* If attached to the endoplasmic reticulum, the ribosomes appear to be lined up along the profiles of the ER strands.

The ribosomes can be isolated from the ergastoplasm, from the rough endoplasmic reticulum or from bacteria. In *Escherichia coli* they have a diameter of 15 nm and consist of two subunits with the sedimentation characteristics 30 S and 50 S corresponding to molecular weights of 0.85 and 1.8 million, respectively (WATSON 1963). The subunits hold 23 S and 16 S ribosomal ribonucleic acid (rRNA). In eukaryotic plants the ribosomes have a higher sedimentation constant of 80 S with 25 S and 16 S rRNA in the subunits, while in eukaryotic animals these figures are 80 S, with 28 S and

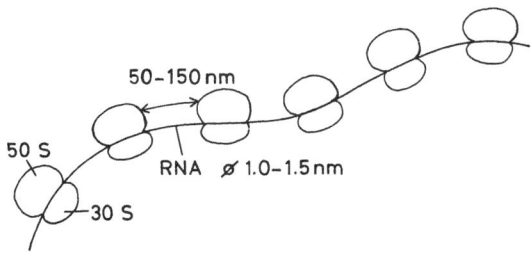

Fig. 16. Polysome. The protein producing organelle consists of a number of ribosomes with their 50 S and 30 S components united by an RNA strand (according to data of GOODMAN and RICH 1963).

18 S rRNA (REISNER *et al.* 1968), indicating that the molecular weight of eukaryotic ribosomal nucleic acids is greater than that in prokaryotes.

If a purified sediment of ribosomes is brought together with appropriate buffers, so-called messenger (mRNA) and transfer (tRNA) ribonucleic acids (which are different from ribosomal rRNA), ATP as an energy source and a mixture of amino acids, they synthesize oligopeptides and polypeptides *in vitro.* This is the proof that ribosomes are involved in protein synthesis. The specific sequence of the amino acids in the peptidic chain is determined by the messenger and the transfer RNA (NIRENBERG 1963).

Individual ribosomes are not capable of achieving the polymerization described, however, a certain number of them acting in cooperation are. This acting group is tied together by the chain of messenger RNA which brings, with its codons (JACOB and MONOD 1961), the necessary genetic information from the nucleus. It seems that the mRNA thread is anchored in the indentation between the 30 S and 50 S subunits of the ribosomes (Fig. 16).

These beaded threads are called *polysomes.* They often assume a spiral or helical shape and were already observed in the early days of electron microscopy by STRUGGER (1957). Although he did not know of their beaded nature, he described them not incorrectly as "cytonemata".

The polysomal thread with its ribosomal beads is not a static but an ephemeric dynamic structure. There is only temporary contact between the ribosome and the ribonucleic strand from which the genetic code is read. Changes of the number per cell of polysomes and free ribosomes are often observed (Linskens and Schrauwen 1968). In *Escherichia coli* it has been found that all the ribosomes which are not integrated within a polysome exist as either free 30 S or 50 S particles. For protein synthesis the 30 S subunit forms an initiator complex with tRNA and mRNA, and only then the 50 S subunit necessary for the translation of amino acids joins the polysomic chain. After its synthetic action the 70 S ribosome falls apart into its subunits (Guthrie and Nomura 1968). It seems that such a ribosomal cycle is a general feature in protein synthesis.

The polysomes are a good example for demonstrating that physiological functions are not performed by individual macromolecules but by complexes of differently high polymeric particles. Only such a cooperative organization can act as an *organelle*.

Separated ribosomes lose the faculty of synthesizing the specific proteins prescribed by the polysomal mRNA (Marcus *et al.* 1967). Similar conditions are valid for the enzymes. Whilst lyoenzymes can act individually in hydrolyzing glucosidic, peptidic, or ester bonds, for the more complicated serial reactions in physiological cycles (Krebs cycle, Calvin cycle, etc.), the catalysts are insoluble desmolases organized twodimensionally, as for instance in the inner membrane of the mitochondria.

3.5.2. Intracellular Translocation and Nutrition

The proteins synthesized by the rough endoplasmic reticulum appear partly —in which proportion is not known—in the enchylema. How these macro-molecules, which are produced on the outer surface, pass through the ER membrane is a problem of its own. As a dilute solution of water-soluble protein molecules, nucleosides and salts, the enchylema has similar properties as blood serum and also the same *nutritional functions*.

This is evident in the case of the nuclear envelope (see 3.1.2.). As we have learned, it is perforated and disintegrates during mitosis; therefore, its function cannot be that of a compartmentation barrier, nor does it represent a protecting sheath for the chromosomes because it is absent during their distribution. Its restoration in the telophase with a complete perinuclear space shows that the enchylema must play a decisive role. In the telophase, inter-phase (John and Lewis 1969) or, as the case may be, in the early prophase of the first meiotic division (Moses and Taylor 1955, Antropova and Bogdanov 1970), the chromosomal DNA material halved by the event of the anaphase, must be reconstituted (Ruch 1966). How the necessary nucleosides and phosphates, amino acids of oligopeptides for DNA and histones are fed to the nucleus is not known. However, it is a plausible hypothesis that these building units are channeled to the perinuclear space by the ER system.

Based on this view, the formation of the nuclear envelope in cell phylogeny was accomplished by the prokaryotes not so much for the better protection of the genetic material in the cell, but rather for the better nutrition of the growing nuclei.

Not only the nucleus but also the chloroplasts may be surrounded with an ER sheath which communicates with the nuclear envelope as demonstrated in the algal groups of Chrysophyceae and Cryptophyceae (GIBBS 1962), Bacillariophyceae, Phaeophyceae, Euglenophyceae, Chrysophyceae and Xanthophyceae (MASSALSKI and LEEDALE 1969). In the archegonium of *Ginkgo biloba* there is even open communication of the periplastidal space with the sheath cisterna (CECCHI FIORDI and MAUGINI 1972) showing that

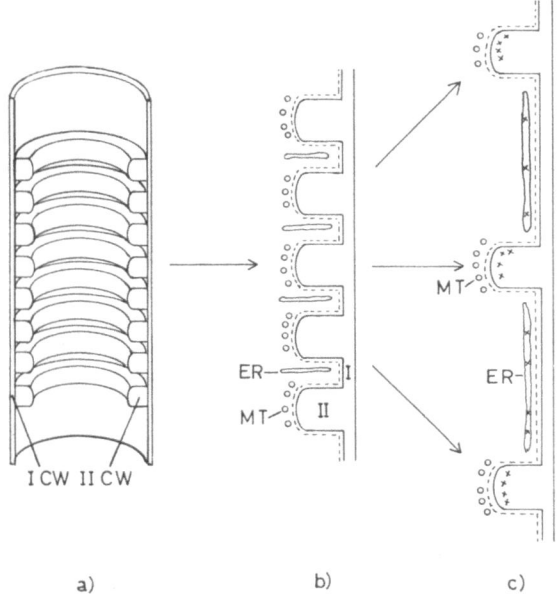

a)   b)   c)

Fig. 17. Differentiation of ring elements in the xylem of wheat seedlings (according to data of PICKETT-HEAPS 1966 and 1967 b). *a*) In the light microscope *I CW* primary cell wall continuous, *II CW* secondary cell wall in form of separated rings. *b*) In the electron microscope ER cisternae between, and microtubules (*MT*) on top of the developing ring. ----- Plasmalemma. *c*) After the elongation growth of the xylem element, the *I* wall is stretched and the ER cisternae lie closely along the wall. × × Label of tritiated glucose fed is found in the ER and in the youngest apposition layer of the rings.

the enchylema of the periplastidal space and that of the ER system are alike and indicating that the outer membrane of the plastidal envelope may be an ER membrane (see 3.3.).

Also the vacuoles of plant cells can be ensheathed by a fenestrated rough ER cisterna as demonstrated by FINERAN (1970, 1972) in the roots of *Avena* seedlings.

A fourth example of the provision of nutritional material for growing parts of a cell by the ER system are the extending ring tracheids in the xylem of germinating seeds (wheat seedlings). The cells for future water conduction produce the ring-shaped thickenings through local cell wall growth (Fig. 17 *a*). In this way the continuous primary wall is covered by an interrupted secondary wall. The whole wall system is coated by the plasmalemma. Opposite the ring profiles of microtubules (see 5.1.) and between the rings, ER elements (Fig. 17 *b*) are visible in the cytoplasm (Pickett-Heaps 1967 b).

When such a young tracheid elongates, the primary wall between the ring thickenings grows considerably in length, so that the rings become separated by many times their width (Fig. 17 *c*). The ER strands which were originally orientated perpendicularly to the cell axis extend parallel to it and now follow the cell wall for long distances in immediate proximity to the plasmalemma without any contact of the two membranes. From the distribution of the cell organelles one might think that the microtubules had something to do with the apposition growth of the ring-shaped secondary wall and the ER elements with the extension growth of the primary wall. However, experiments with tritium labelled glucose showed that the endoplasmic reticulum only distributes labelled molecules fed to the seedling (Pickett-Heaps 1966). The label appears in the ER cisternae, which act as distributors from which the label is forwarded to the tops of the ring profiles as indicated in Fig. 17 *c*.

As the matrix for the extension growth of the primary wall is furnished by Golgi vesicles (see 2.3.3.2.), it is not unexpected that the label which characterizes ER material does not appear in the primary wall. On the other hand, the fact that the cellulosic elementary fibrils which constitute the ring-shaped secondary wall are labelled indicates that the glucose necessary for extracellular cellulose synthesis is supplied by the endoplasmic reticulum. The necessary nitrogen compounds for the growth of the protoplast in extending cells must be distributed in a similar way by the endoplasmic reticulum.

If the ER cisternae correlated to the growth phenomena described are smooth, the nutritional material must be translocated from the rough region to the local places of growth. All smooth cisternae must receive the polypeptides of their enchylema from rough ER elements so that a further function of the ER system is *intracellular translocation*.

This transportation of micro- (and macro-?) molecules is much faster than molecular dislocation by diffusion. However, no clues are available for a mechanistic explanation of such energy consuming translocation. This problem is just as enigmatic as the intracellular transport of Golgi or ER vesicles where we do not know by which principles these vehicles reach their goal.

As ER strands extend through the plasmodesmata from cell to cell there is also a possibility of *intercellular translocation* of molecules by-passing the barriers of the plasmalemma or other compartmental membranes. Such transportation certainly exists through the cell plate during mitosis (Frey-Wyssling et al. 1964), but it is difficult to say how long it is operative after cell division (cp. Plate VII, Fig. Q).

3.5.3. Formation of Metabolic Centres

The ER system is not only responsible for the compartmentation which separates the enchylema from the groundplasm but also for the creation of a number of vesicular compartments with special physiological functions.

3.5.3.1. *Spherosomes* (Plate VII, Fig. R_1)

Spherosomes are well defined organelles in plant cells which were described as early as 1880 by HANSTEIN. At that time they were called "microsomes" which term was later altered to *spherosomes* (PERNER 1953). As implied by their name, they are spherical. They stain with lipophilic dyes and are usually larger than 1 μm in diameter. As their refractive index is appreciably higher than that of the groundplasm, they are readily visible in the light microscope.

They are produced by the ER system (FREY-WYSSLING *et al.* 1963). Osmiophilic material is accumulated in the end lobe of a reticulum strand.

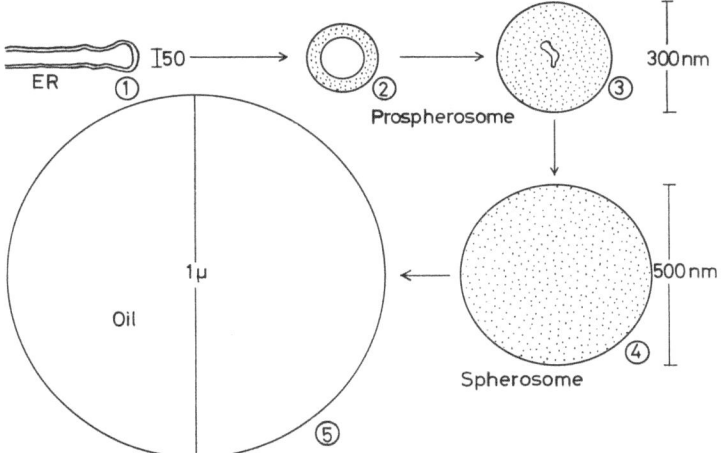

Fig. 18. Ontogeny of spherosomes and oil droplets in plants. (1) ER strand, (2) prospherosome, (3) ER membrane undergoes lipophanerosis, (4) spherosome, (5) oil droplet (according to data of FREY-WYSSLING *et al.* 1963 and SCHWARZENBACH 1971).

Then a small body coated by an ER membrane buds off and grows in a short time to a diameter of 100–150 nm (Fig. 18). Such bodies traceable in the ultramicroscope have been called "prospherosomes" (JAROSCH 1961). Later they assume light-microscopic dimensions when their membrane which originally is a trilamellate unit membrane loses this aspect. The stratum between the blackened osmiophilic lines (Plate VII, Fig. R_2) inflates and fills the spherosome centripetally. In the center an unstained area is visible which is delimitated by a black line (Fig. R_2, p. 41).

As a result of this ontogeny the spherosome, and the finally resulting oil droplet (Fig. 18), are no longer coated by a unit membrane but only by a single black lamella which represents the outer osmiophilic stratum of the original ER membrane (SCHWARZENBACH 1971).

The function of the spherosomes has been investigated by testing their enzymatic capacities. So far, three objects have been examined in this way: the spherosomes in the scales of the onion bulb (*Allium cepa*), in the coleorhiza of corn seedlings (*Zea mays*) and in the endosperm and embryo of germinating tobacco seeds (*Nicotiana tabacum*, MATILE 1969 d).

In the first case (*Allium*), the epidermal cells of the onion scales were tested histochemically (Walek-Czernecka 1965) whereby acid phosphatase, esterase, β-glucuronidase, β-galactosidase, β-glucosidase, lipase, arylsulphatase, as well as acid DNase were found. As the different cell organelles may interfere with each other in such experiments it is not certain whether all the enzymes mentioned are really located inside the spherosomes. Drawert (1953) showed for instance that indophenol blue generated by the cytochrome oxidase of mitochondria is stored in the lipids of the spherosomes.

A more reliable method is the isolation of the spherosomes in an ultracentrifuge whereby the influence of other cell constituents can be eliminated. In this way it is found that the spherosome fraction of tobacco seedlings holds protease, lipase, esterase, acid phosphatase, RNase and DNase (Matile and Spichiger 1968), whilst that of the maize coleorhiza contains only lipase (Matile 1971).

The only enzyme common to all spherosomes investigated is lipase. This speaks for the view that the spherosomes are centers for oil synthesis and oil storage (Frey-Wyssling et al. 1963). Sorokin (1967) denies any relation between spherosomes and oil droplets in storage tissues because the former remain after lipid extraction whereas the oil bodies leave not even a membrane. Yet, such statements cannot invalidate their established ontogeny. The insolubility of spherosomes in apolar solvents is due to their protein content which vanishes during the accumulation of triglycerides.

Since the so-called essential fatty acids (unsaturated oleic, linolic, linolenic, arachidonic, etc. acids) cannot be synthesized by animal and human cells but only by plant cells, it seems not unexpected that for this activity a special organelle is provided in plants. The unusually high refractivity of the spherosomes might be due to their unsaturated lipid compounds. Based on the function of synthesizing vegetable oils, it has been proposed to replace the morphological term "spherosome" by the more functional term "oleosome".

The presence of other enzymes than those for fat metabolism is an indication of additional activities of these ER-born organelles. It is characteristic that only in the coleorhiza with its moribund tissues lipase alone seems to be active, while in the endosperm of germinating seeds, a clear-cut distinction between spherosomes and lysosomes is difficult.

3.5.3.2. *Lysosomes*

Lysosomes (De Duve 1959) are compartments of the cell which hold lytic enzymes. They are autophagic vacuoles (De Duve and Wattiaux 1966) with the necessary hydrolases for the intracellular turnover of proteins and nucleic acids. Originally they were discovered in the liver, the tissue of which can be autolyzed by the contents of the isolated lysosomes for which reason they have been called "suicide bags" (De Duve 1963). Their membrane which is a unit membrane prevents such destructive action in the living cell where the lysosome functions as an organelle for intracellular digestion.

The origin of the lysosomes is the same as that of the spherosomes. They are produced by ER strands which give off small vesicles. The same oxidoreductases known as desmolases on ER membranes (cytochrome c, NAD oxidoreductase and NADH DIP diaphorase) are found on lysosomal membranes (MATILE and WIEMKEN 1967, MATILE 1969 b). The enzymes of the vesicles may originally derive from rough regions of the endoplasmic reticulum. This is especially understandable if there is a sheath of rough ER around the lysosome (FINERAN 1972) as mentioned in 3.5.2. However, when the organelles increase in size and change their specific hydrolytic activity during cell differentiation, the lysosomes seem to be capable of synthesizing their own enzymes. On the surface of the yeast vacuole which functions as a lysosome, ribosome-like particles have been found (WIEMKEN 1969). Whether there is an autonomous protein synthesis taking place is currently being tested with isolated lysosomes to which various substrates are offered for hydrolysis (MATILE 1972).

As the hydrolases are lyoenzymes they are probably dissolved in the enchylema of the lysosomes and not adsorbed to the lysosomal membrane which is involved in transmitting or producing these enzymes.

The vesicles described are called *primary lysosomes*. These may fuse with each other or with larger vacuoles in the cell and form large *secondary lysosomes* with phagocytotic capacities. For instance they may take in degenerating mitochondria and digest them.

Lysosomes are not only characteristic of animal cells, but also of plant cells (MATILE 1969 a). In the hyphae of the fungus *Neurospora crassa* special particles were found which contain proteolytic enzymes (MATILE 1964); therefore, they were put on a par with primary lysosomes. These particles play an important role in the extracellular digestion of protein. Although *Neurospora* is a nitrogen-autotrophic organism which can live with nitrate as a N-source, it may also be fed with organic N-compounds. If soluble proteins such as gelatine or haemoglobin are added to the culture medium, these compounds are hydrolyzed outside the cell and the resulting amino acids are resorbed.

The mechanism of this extracellular proteolysis could be disclosed by MATILE (1965/1967). As soon as protein appears outside the cell, the production of primary lysosomes is increased. These migrate to the cell surface and gather next to the plasmalemma where they are extruded as described in Fig. 14. The special feature of this phenomenon is that the lysosomal unit membrane which derives from the ER membrane does not fuse with the plasmalemma, but is still intact when the particle appears in the extracellular space between plasmalemma and cell wall. Only subsequently this membrane is dissolved and the proteases are set free. They pass through the holopermeable cell wall by diffusion and digest the extracellular proteinic substratum.

These physiological events raise a number of questions which show that discoveries in biology will never yield final results but, on the contrary, create new unanswered problems: e.g. how does the nucleus of *Neurospora* get the information that extracellular digestible proteins have appeared (Fig. 19) so that instead of nitrate, amino acids may be resorbed? How does the nucleus transfer its information to the endoplasmic reticulum? How does this organelle "know" that in this case the production of primary lysosomes must be intensified? Which force translocates these particles to the surface of the cell and how does the plasmalemma "realize" that just these and no other cell constituents must be secreted into the extracellular space? And the last question, why is the lysosomal membrane preserved as long as it is transported through the groundplasm and the plasmalemma, whereas it decays in the extracellular space?

From corn seedlings two types of lysosomes, heavy ones and light ones, have been isolated (Matile 1968 a). Both hold protease, carboxypeptidase, DNase, RNase, phosphatase and various esterases. However, there is a

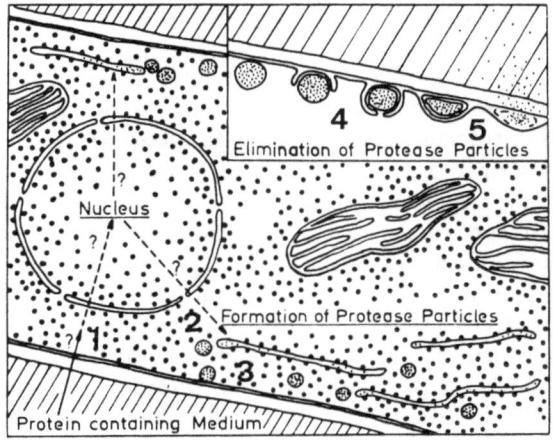

Fig. 19. Diagrammatic representation of the events when *Neurospora* is fed with protein instead of inorganic NO_3 (courtesy of Ph. Matile, from Abh. Dtsch. Akad. Wiss. Berlin, Akademie-Verlag, 1967). (1) The nucleus is informed of the presence of extracellular protein, (2) the ER is instructed to produce lysosomes in excess, (3) the lysosomes line up under the PL, (4) they are extruded by the PL (see inset), (5) the ER membrane does not fuse with the PL but is dissolved outside the cell and releases the proteases of the lysosome into the extracellular space.

difference in that the light lysosomes contain in addition transaminases and the heavy lysosomes oxidoreductases known to occur also in the membranes of the reticulum.

The light lysosomes with their less dense contents resemble vacuoles; and if they fuse the secondary lysosomes formed can no longer be distinguished from vacuoles.

3.5.3.3. *Vacuoles (Tonoplast)*

The large central vacuole in plant cells is delimited from the surrounding plasmic layer by a membrane termed *tonoplast* by De Vries (1885) long before it could be portrayed as a unit membrane in the electron microscope.

The vacuole contains the cell sap which is a dilute solution of salts, organic acids, sugars, soluble tannins, anthocyanins and other polyphenolic compounds. The central vacuole originates in meristematic cells through the fusion of small vacuoles, which in very young cells can be traced in the

electron microscope down to so-called *provacuoles* invisible in the light microscope. The provacuoles on their part are produced by inflations of ER strands.
Thus the vacuolar membranes and the tonoplast are ER membranes (BUVAT
1962). Freeze etching discloses particles in the tonoplast which may be
involved in enzyme synthesis (Plate VIII, Fig. *S*).

The functions of the central vacuole in plant cells are manifold. In the
first place they are involved in *osmoregulation* which guarantees turgescence
by an appropriate turgor pressure T established as the difference between
the theoretical osmotic pressure O of the cell sap and the suction pressure S
(tendency to take in water) of the cell: $T = O - S$.

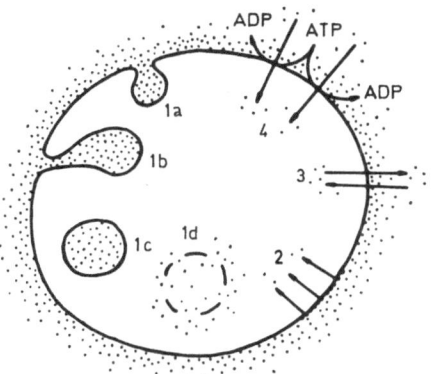

Fig. 20. Activities associated with the yeast vacuole. (1) The intake of plasmic material
into the lysosomal compartment by means of endocytotic-like activity of the vacuolar
membrane (*a–d*) has been observed in freeze-etchings, (2) the vacuolar membrane is involved
in the synthesis of lysosomal enzymes, (3, 4) the mutual interaction between the vacuole and
the cytoplasm requires the transport of micromolecules through the vacuolar membrane
[courtesy of PH. MATILE, from Prospects of yeast cytology, Antonie van Leeuwenhoek 35,
suppl. 59 (1969 c)].

A second function is based on the fact that in ageing cells the vacuole
is a large extraplasmic compartment the contents of which are separated from
the metabolizing protoplast by the tonoplast. As a consequence all kinds of
metabolites, which in animal cells would be eliminated into the extracellular
space are excreted into the vacuole. The list of such metabolites covers a
wide range from alkaloids (e.g. nicotine, caffeine) over polyphenols such as
anthocyanins, flavones and tannins of both the gallic acid (C_7) and catecholic
(C_{15}) types to steroids (saponins). A prerequisite for this *excretion* is the
water solubility of the compounds to be eliminated. If these are insoluble
in water, they are converted into glucosides through combination with sugar
molecules (FREY-WYSSLING 1942). As an example the insoluble anthocyanidins
are transformed into the soluble anthocyanins in this way.

A third function consists in the *storage* of assimilates such as sugars and
proteins. The accumulation of proteins in seeds occurs in so-called *aleurone
vacuoles* which are filled with albumins, globulins and the hexaphosphate
phytin, whereupon the vacuole is dehydrated so that solid aleurone grains

are formed. During germination these grains are rehydrated and reconverted into vacuoles. Matile (1968 b) has shown that these vacuoles contain protease, RNase, but no DNase, β-amylase, α-glucosidase, phosphatase and esterase. Therefore, aleurone vacuoles are capable of lysosomal activity, not for a direct autophagic turnover, it is true, but for the digestion of the stored protein and phytin when the seed germinates.

Thus, a fourth function concerns *digestive activities*. The vacuole of the yeast cell is such a typical lysosome. It is uncertain whether it derives from the endoplasmic reticulum because, during the process of budding, small vacuoles produced by fragmentation of the mother vacuole wander into the daughter cell and there constitute a new central vacuole. In contrast to other plant cells there seems to be a continuity of the vacuolar system in the ontogeny of yeast. The mature vacuole displays striking phases of autophagic turnover (Plate VIII, Fig. *T*). During cell growth and differentiation plasmic material is introduced into the vacuole (Matile 1969 c) and digested (Figs. 20/1 *a*–1 *d*). The necessary enzymes are probably provided from inside the tonoplast (Plate VIII, Fig. *S;* Fig. 20/2). The transport of micro-molecules (e.g. sugars into the vacuole and of the digestive products (amino acids, etc.) out of the vacuole is indicated by the arrows in Fig. 20/3. The necessary energy for these translocations is furnished by plasmic ATP (Fig. 20/4).

The energy source is not only necessary for the transport of molecules through the tonoplast but also for the maintenance of its labile structure. In this respect it seems important that the inside of the cytomembrane which is in contact with the vacuolar lyoenzymes is not digested by the available proteases and phosphatases. However, the situation changes completely when tonoplast bound digestive vesicles are produced by invagination (Fig. *T*, Fig. 20/1 *a*–1 *c*). In this case the inner surface of the tonoplast which previously resisted digestion becomes the outer surface of the vesicle and is readily attacked (Fig. 20/1 *d*). This pleads for the view that there is no energy source in the vacuoles so that the ultrastructure of the membrane maintained in contact with the groundplasm by constant energy consumption decays easily when separated from the source of ATP.

There are other vacuolar systems with lysosomal functions. In the latex of *Hevea brasiliensis* which is a highly diluted cytoplasm, there are slightly yellowish vacuoles called *lutoids*. They can easily be separated from the latex serum after the rubber is skimmed off and prove to contain protease, phosphatase, RNase, DNase, β-galactosidase and α-glucosidase (Pujarniscle 1968). Therefore, these vacuoles must also be classified as lysosomes.

Plate VIII. Fig. *S.* Tonoplast membrane of yeast, isolated vacuoles frozen, broken and then deep-etched (cp. Fig. 2 c); a) outer surface of the membrane smooth, b) inner fracture face, rough with particles (enzyme complexes?), c) regions free of particles, ×45,000 (courtesy of F. Kopp). Fig. *T.* Lysosomal activity of yeast vacuole (cp. Fig. 20). *CW* cell wall, *V* vacuole with invaginations and resorbed vesicles, *N* part of nuclear envelope, ×20,000 (courtesy of Ph. Matile).

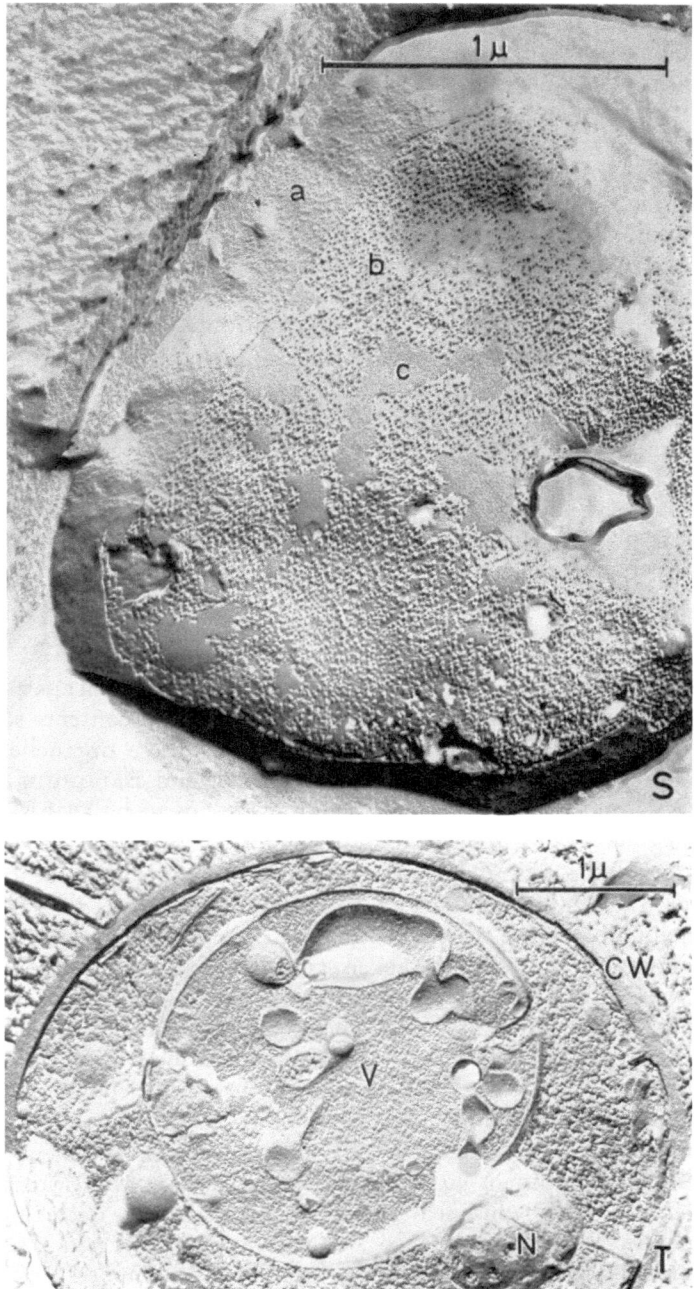

Plate VIII.

Based on the experience with the examples mentioned, it seems difficult to differentiate between spherosomes (oleosomes), lysosomes and vacuoles. Especially in the early stages of their ontogeny when they split off from the ER system as vesicles, their future behavior cannot be predicted. To characterize them they must be isolated in a centrifuge and then tested for their enzymatic capacities. Marker enzymes for *lysosomes* are protease, phosphatase and a fair activity of RNase, the last of which is also symptomatic for the rough endoplasmic reticulum. In addition to such enzymes *spherosomes* contain lipase as a marker. In the *provacuole* enzymes may be lacking (Matile 1969 b) but appear later when the turnover of the metabolizing cell comes into operation. At this stage the function of the vacuoles as digesting organelles of the plant cell becomes manifest.

This lysosomal activity is probably the most important task of the small plant vacuoles before they fuse into a cell-filling central vacuole. As a matter of fact, cell saps (must, incl. new wine) as well as saps of sugarcane and sugar beet no longer display distinct proteolytic activity. Of course there are exceptions such as pine apple sap or papaya juice the capacity of which for hydrolyzing proteins is known. The proteolytic enzyme papain of the milky juice of papaya is produced in the vacuoles of the laticifers in the fruit.

3.5.3.4. *Peroxisomes*

A further membrane bound organelle is the *peroxisome* (Tolbert et al. 1969). These particles have solid sometimes even crystalline contents so that they were called "microbodies" before the function of these organelles was known. They were discovered in the liver (De Duve and Baudhuin 1966) in the same way as the lysosomes. Their membrane seems to be an ER membrane because the microbodies are said to derive from the reticulum.

The peroxisomes have a similar density as the mitochondria (mitochondria 1.19, proplastids 1.23, glyoxisomes 1.25 g/cc, Breidenbach et al. 1968) so that it is difficult to isolate them in a pure fraction (Parish 1971 b). Successful fractionation shows that the enzyme catalase which decomposes hydrogen peroxide ($H_2O_2 \rightarrow H_2O + O$) is not localized in the mitochondria but in the peroxisomes. Therefore, catalase is the marker enzyme for peroxisomes by which they can be distinguished from lysosomes and other organelles. They seem to be a general organelle found in many types of animal and plant cells (Matile 1969 b).

Other enzymes may be associated with catalase. At the moment three types of peroxisomes are known in plant cells (Parish 1971 a): *Glyoxisomes* (Breidenbach et al. 1968) which catalyze the glyoxylic acid cycle in connection with the transformation of seed fats into sugars, *glycolisomes* with glycolic acid oxidase and *uroxisomes* with uricase that decomposes uric acid.

It is easily seen that this armory of enzymes is fundamentally different from the hydrolases met with in lysosomes. However, the cooperation and the physiological significance of the peroxisomes is not yet as clear as that of the lysosomal apparatus.

3.5.4. Review

A review of the functions attributed to the endoplasmic reticulum is given in Table 3.

Table 3. *Functions of the Endoplasmic Reticulum and its Derivatives*

Anabolism	synthesis of proteins
	synthesis of lipids
Translocation	intracellular transport of proteins
	intercellular transport through plasmodesmata
Accumulation	protein and hexaphosphate (phytin) in aleurone grains
Katabolism	intracellular digestion in lysosomes (turnover)
Elimination	exocytosis of exoenzymes (see 3.5.3.2. and Fig. 19)

3.6. Derivatives of the ER Membrane

The formation of spherosomes, lysosomes, vacuoles or peroxisomes out of ER elements necessitates a transformation of their membranes. As already pointed out, it is evident that not all enzymes in these organelles derive from the rough endoplasmic reticulum but that they must synthesize their characteristic catalysts themselves. As the membranes are the site of this synthesis which occurs with the aid of plasmic ATP located on their outer side, they must differ slightly from each other in the various derivatives and from the original ER membrane. The difference is not as striking as in the case of myelin lamellae which can easily be distinguished from the prototype of the plasmalemma from which they derive, but their enzyme complexes must be physiologically different.

Since the lipids synthesized in spherosomes appear in the middle stratum of their trilamellated unit membrane (Fig. 18; Plate VII, Fig. R_2), the metamorphosis postulated is even visible. In this case the accumulation of glycerides is such that the unit membrane character is lost and, in the electron microscope, the resulting oil droplets are coated on their profiles by a simple black line (SCHWARZENBACH 1971). In this respect there is a considerable difference to the lamellae of the myelin sheath where the widths of clear lipid strata (phospholipids and steroids) and black protein strata are similar.

3.7. Survey of Membrane Metamorphosis

Under 3.4. it has been shown how in lower organisms ER membranes can be transformed into Golgi membranes which are homologous and analogous to the plasmalemma. The change ER membrane → mature Golgi membrane necessitates considerable growth and metamorphosis during the shift of the Golgi cisternae from the receiving to the secreting pole of the dictyosome (Fig. 7). A direct incorporation of ER membranes into the plasmalemma is not observed in higher plants. It is true that in nectaries the ER elements which contain the nectar are not extruded as membrane-bound vesicles (as in Fig. 14) but simply emptied into the extracellular space. (Fahn and Rachmilevitz 1970). However, this does not prove that these ER membranes are incorporated into the plasmalemma, because there is no increase in width or area of the cell membrane.

Table 4. *Survey and Interrelation of Cytomembranes*
PL plasmalemma, ER endoplasmic reticulum

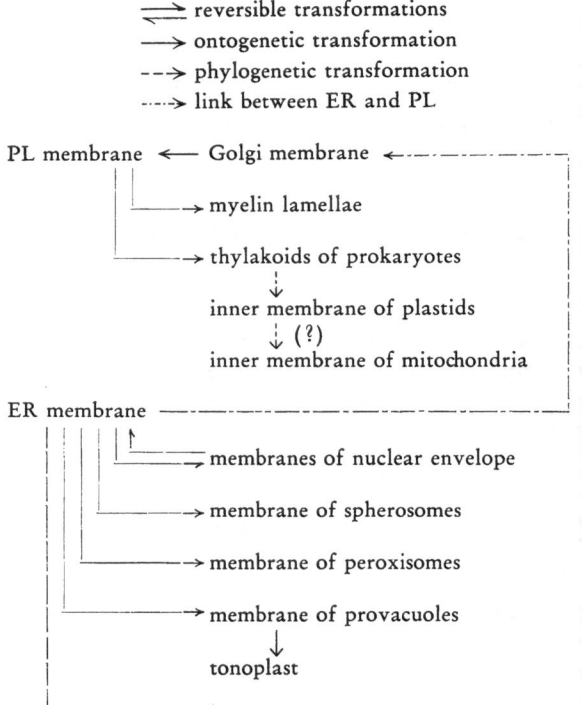

The transformation of membranes from one type to another type has been called "flux of the membranes" (in German: Membranfluß) by Schnepf (1969). This term is appropriate in so far as a flow is directional and, as a rule, not reversible. However, it implies a liquid state of the membranes which is wrong. Membrane-bound vesicles do not fuse in the way of water droplets in the atmosphere or oil droplets in an emulsion. In the centrifuge the biomembranes retain their individuality. Certainly, ER cisternae round up to form "microsomes" when isolated, and the inner membrane of mitochondria unfolds in the phenomenon of ballooning, but they never merge. Therefore, the fusion of Golgi vesicles or provacuoles in the cell and the integration of Golgi membranes into the plasmalemma cannot be simple physical processes but are physiological acts supported by energy transfer and restricted to homologous membranes.

In this respect we can distinguish two families of membranes in higher plants: the PL membranes and the ER membranes. It has been shown that ER membranes can evolve into PL membranes. However, this is a complicated maturation process which in higher plants has not yet been observed as clearly as in *Pythium* (GROVE *et al.* 1970), *Glaucocystis* or *Vacuolaria* (SCHNEPF and KOCH 1966 a, b).

Table 4 gives a survey of the interrelations of the biomembranes under consideration.

The plasmalemma can be produced or regenerated by the Golgi apparatus. In special cases it metamorphoses into myelin lamellae in nerves or thylakoids in prokaryotes. A possible phylogenetic connection between the inner membrane of plastids or mitochondria is indicated by a dotted arrow.

The family of organelles with ER membranes is larger, because the endoplasmic reticulum produces highly specialized cell compartments. The link between the ER and the PL membranes is the Golgi apparatus which is capable of transforming ER membrane into the more complicated PL membrane.

A special feature of Table 4 is that the transformations indicated by arrows are uni-directional. There is but one exception in the case of the nuclear envelope which can decay into ER elements and may be reconstituted from them. For the rest the metamorphoses seem to be irreversible. A highly specialized membrane cannot return to the structure of its more versatile ontogenetic or phylogenetic precursor.

4. Undulipodia and Centrioles

4.1. Ultrastructure

4.1.1. Flagella and Cilia

Flagella and cilia have similar diameters (around 0.2 µm) and the same ultrastructure. This fact justifies the common term *undulipodium* for the two types of motile organelles. Their only difference is in length (cilia 5–10 µm, flagella up to or over 150 µm) and in number per cell. The flagellates carry one flagellum or two unequal flagella (Heterokontae), while the ciliates are equipped with a great number of cilia.

A section across the cylindrical shaft of these organelles invariably discloses eleven fibrils embedded in an electron-transparent matrix the membrane of which is continuous with the plasmalemma. Nine of the eleven strands are arranged in a ring around two central fibrils. The outer fibrils are doublets of microtubules termed A and B while the central strands are single tubules. The A tubules carry hornlike arms (ALLEN 1968, see Figs. 21 *b* and 23). The standard sizes of the flagellar elements are given in Fig. 21 (FAWCETT 1961).

One of the most fascinating discoveries of electron microscopy disclosed that all undulipodia, from the flagellates and ciliates to the lower plants with antherozoids (MANTON 1956) and through all phyla of the animal kingdom up to the mammalian including human sperms and ciliated epithelia (Fawcett and Porter 1953) are of the 9 + 2 stranded type.

Only the flagella of prokaryotic organisms are of a simpler ultrastructure. In bacteria the shaft of the flagellum consists of merely one microtubule. Its diameter measures 20 nm in *Pseudomonas fluorescens*. Morphologically it thus corresponds to one of the central tubules in Fig. 21 (Ringo 1967 a). Why in eukaryotes this strand had to be doubled and equipped with nine peripheral fibrils nobody knows. As single tubules are functional in bacteria,

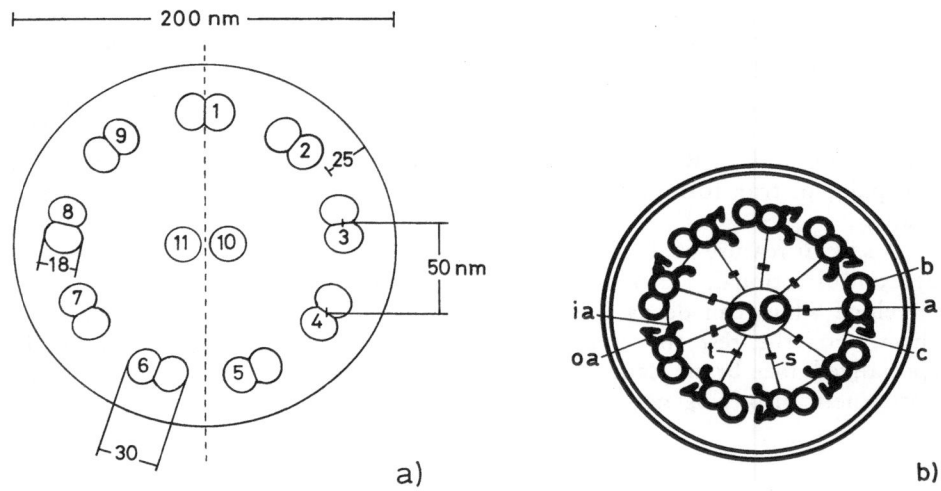

Fig. 21. *a*) Diagrammatic cross section of an undulipodium. 1 ... 11 number of fibrils, ------ plane of symmetry (according to data of Fawcett 1961). *b*) Ultrastructure disclosed by Markham rotation technique (from Allen 1968). *a* tubule A, *b* tubule B, *c* connection between doublets, *oa* outer arm, *ia* inner arm of tubule A., *s* spoke, *t* thickened region along spoke (filament).

an assembly of such fibrils will certainly produce more vigorous motility, but how the eleven strands cooperate is still a mystery. Since only the basal body of the undulipodium (see 4.1.3.) is provided with energy by means of ATP and the outer fibrils (1–9 in Fig. 21 *a*) are devoid of ATPase (Gibbons 1967), the shaft with its 9 outer fibrils seems to play a merely passive role comparable to a whip which is handled at its base (see however 4.4.1.).

Until dynamics find an explanation for the advantage of a model with 9 + 2 fibrils above models with another number of peripheral strands or without them, the established ultrastructure must be accepted as a given pattern.

Two conclusions can be drawn from these findings:

1. that all eukaryotes seem to be of monophyletic origin, and

2. that during half a billion years no mutational evolution of the flagellum structure has occurred! Whether any such change proved lethal is not known, but what we must recognize is an amazing conservatism which clings, so to speak, indefinitely to an established and approved structure.

High resolution electron microscopy shows that the tubules are built of globular subunits. These are protein molecules with a molecular weight of about 40,000. They have diameters of 4–5 nm and are arranged in straight longitudinal rows. Sections show 13 such particles per tubule. In doublet strands adjacent tubular walls merge so that the cross-section of the double tubules is slightly elliptical (Fig. 22).

In bacterial flagella the tubule does not show 13, but 5, 6 or 8 spherical subunits on the cross section (RINGO 1967 a). Also the A tubule of the centriole in human leucocytes holds only 11 instead of 13 subunits (Ross 1968).

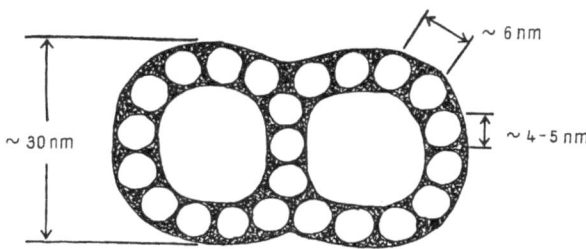

Fig. 22. Diagrammatic interpretation of the arrangement of subunits in a flagellar doublet seen in cross section. Approximate dimensions are given for the diameter of the fibrils, center-to-center spacing of subunits, and diameter of subunits [D. L. RINGO, from Ultrastructure Res. **17**, 266 (1967 a)].

4.1.2. Terminology

The flagellar strands are referred to as fibres and the tubules as subfibres. This terminology dates from the time before the spherical subunits of the tubules were discovered. These globular particles arranged in linear array form beaded chains. Such a chain does not display the pronounced anisotropy which is characteristic of fibres with extended cateniform molecules assembled in a chain lattice. As the term "fibre" includes not only the morphological concept of a filiform linear shape, but also the mechanical property of a certain tensile strength, tubules made up of globular macromolecules should not be called fibres or "subfibres" but *fibrils*.

Table 5. *Comparative Size of Flagellar and Fibrillar Constituents*

flagellum	– – –	fibrils	———	microtubules	———	beaded chains
⌀ 200 nm		⌀ 30 nm		⌀ 18 nm		(spherical units)
						⌀ 4.5 nm
cell wall of	– – –	fibrils	—·—·—	microfibrils	———	elementary fibrils
plant fibres		⌀ 400 nm		⌀ 10–20 nm		(chain lattice)
10 µm thick						⌀ 3.5 nm

In Table 5 the sizes of the fibrillar elements in flagella are compared with those of cellulose in the cell wall of higher plants. There, extended chain molecules of cellulose crystallize into elementary fibrils of 3.5 nm diameter; these assemble into microfibrils which associate and form fibrils visible in the light microscope. In cotton hairs such fibrils measure 0.4 µm in diameter.

It is interesting that the flagellar microtubules have the same dimensions as microfibrils and their spherical units of the beaded chains a similar diameter to cellulosic elementary fibrils. Although separated cellulose chain molecules with a width of 0.8 nm can be obtained by artificial disintegration of the hydrogen bonds in elementary fibrils, they do not occur in living cells, because cellulose synthesis proceeds by concomitant polymerization of glucose and crystallization of the nascent polyglucosan chains. Therefore, 3.5 nm strands are the smallest cellulose units met with in the cell of higher plants (Frey-Wyssling 1969).

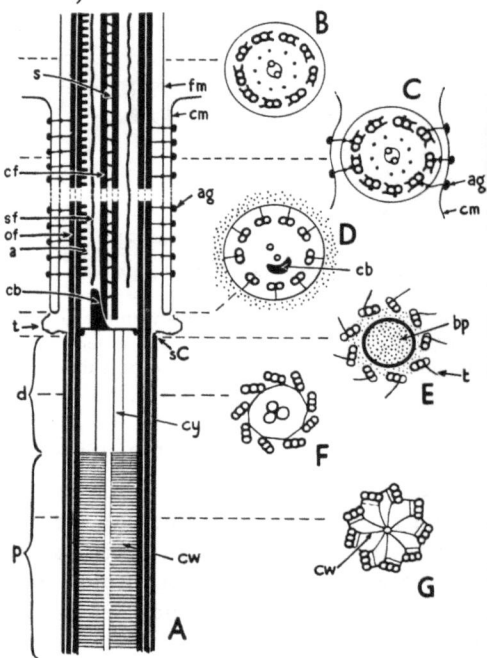

Fig. 23. Diagram of cilium and basal body of *Pseudotrichonympha* [I. R. Gibbons and A. V. Grimstone, from J. Biophys. Biochim. Cytol. **7**, 697 (1960)]. *A*) Longitudinal section, *B–G*) cross sections at the levels indicated, *a* arms of fibrillar doublets, *ag* anchor granule, *bp* basal plate, *cb* crescent body, *cf* central fibril, *cm* plasmalemma, *cw* cartwheel structure, *cy* cylinders, *d* distal region of basal body, *fm* flagellar membrane, *of* outer fibril, *p* proximal region of basal body, *s* central sheath, *sC* distal end of basal fibril C, *sf* secondary fibril, *t* transitional fibril.

As a consequence, building units with dimensions around 4 nm seem to play an important role as structural elements in macromolecular morphology (Table 5).

4.1.3. Basal Body

Every undulipodium roots in a special anchorage called *basal body*. This organelle also has a ninefold symmetry. The nine peripheral tubular doublets of the flagellum continue into the basal body while the two central tubules end above the basal plate between the flagellar shaft and the basal body (Fig. 23 *D, E*).

In the basal body the flagellar fibrils appear as triplets in that a third tubule is added to the doublet. In the lower proximal part of the basal body the nine triplets are united by spokes which run to a kind of central hub. This makes the cross-section look like a cartwheel (Fig. 23 G). This basic structure shows the triplets slightly tilted so that, especially in more distal sections across the basal body where the spokes are lacking, the "wheel" is described as a pinwheel or a cogwheel. In *Pseudotrichonympha* the tilt occurs in a clockwise direction. Although there are a very great number of cilia in this organism, the enantiomorphic anticlockwise arrangement has not been found (GIBBONS and GRIMSTONE 1960).

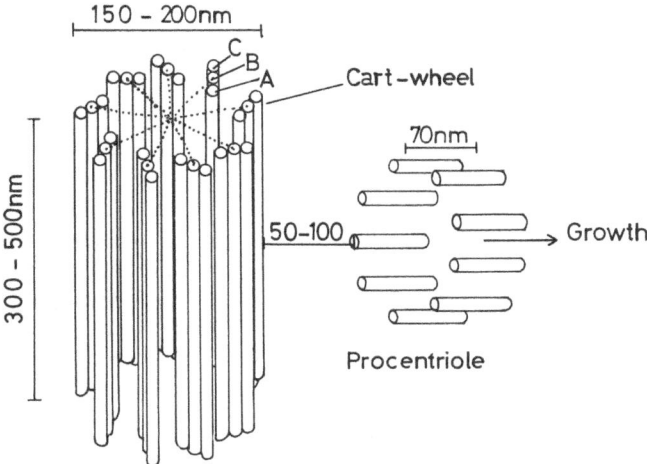

Fig. 24. Diplosome formation. Centriole with triplet tubules *A*, *B*, *C*, produces a procentriole at a distance of 50–100 nm (drawn according to data of FULTON 1971).

The innermost tubule of the triplet corresponds to *A*, the middle one to *B* and the outermost is indicated by *C*. While the tubules *A* and *B* are continuous with the flagellar doublets, *C* is restricted to the basal body (Fig. 23).

4.1.4. Centrioles

Centrioles constitute the centres from which, in mitosis, the spindle fibres radiate as polar asters. At this stage the center consists of two separate centrioles which form a *diplosome*.

In the electron microscope the diplosome appears as an assembly of short microtubules arranged in a zone of groundplasm called *centrosphere* which is exempt of ribosomes, endoplasmic reticulum, mitochondria and other cell organelles. Each centriole consists of nine tubular triplets, 0.3 to 0.5 μm long (BERNHARD and DE HARVEN 1958), in circular arrangement (Fig. 24). A cross-section of this system shows exactly the same pinwheel as a basal body. As will be seen there is not only this morphological coincidence, but also an ontogenetic reason, why centrioles and basal bodies must be discussed together.

The second centriole of the diplosome is always arranged at right angles to the axis of the first. The two bodies are clearly separated and never touch each other (Fig. 24).

4.2. Ontogeny

4.2.1. Formation of Flagella and Photoreceptor Organelles

When a cell produces a flagellum it can be seen how a basal body develops from a centriolar pinwheel and how then the basal body produces the flagellum with its characteristic structure of 9 + 2 strands. Thus there is an ontogenetic development:

centriole → basal body → flagellum.

If, as observed in *Hydra*, diplosomes are involved in the formation of a flagellum, one of the centrioles follows the ontogenetic line indicated, while the other stays unaltered in its position perpendicular to the flagellar axis. However, in other cases it may become the centre for other differentiations. This occurs in phytoflagellates where the second centriole gives rise to the development of a photoreceptor organelle. In *Chromulina* it produces an inflated ciliary organelle against which vesicles with a red carotenoid pigment are aligned (Fauré-Fremiet 1958). The system represents the eye spot of *Chromulina* with which it regulates its phototactic movements. In this way both the flagellum and the photosensitive cilium emerge from basal bodies orientated at right angles to each other (Fig. 25 a).

The outer segment of the retinal rods in the eye of vertebrates represents a similar ciliar differentiation (Fig. 25 b) with a basal body, nine doublet fibrils and pigment-holding flat vesicles (Fauré-Fremiet 1958).

4.2.2. Hispid Flagella

A special feature is represented by the hispid flagella (Fawcett 1961) which are found in the heterokontan green algae, sperms of brown algae, Chrysophyceae, Xanthophyceae, dinoflagellates, cryptomonads and aquatic fungi. In the heterokontan algal group, it is the larger of the two flagella which is hairy. These hairs or *mastigonemata* (Greek *mastix* = whip, *nema* = thread) are arranged along one or two longitudinal lines parallel to the fibrillar doublets of the flagellum; it was originally thought that they were outgrowths of the undulipodium. However, their formation is more complicated.

In brown algae (Bouck 1969) and in Chrysophyceae (Leedale et al. 1970), the hairs are produced as tubules in the perinuclear space and other regions of the endoplasmic reticulum. The tubules have a diameter of 17–19 nm (in *Fucus* and *Ascophyllum*) and are thus more slender than cytoplasmic microtubules with diameters of 20–25 nm (see 5.1. and Fig. 28 a). Large vesicles containing these tubules are detached from the ER system (Plate IX, Fig. W), wander to the cell surface and undergo exocytosis next to the originally smooth flagellum. Outside the cell the tubules are arranged along the flagellum and fixed to it.

The ontogeny of these mastigonemata reminds one of the secretion of scales by naked flagellates (see 2.3.5. and Plate VI, Fig. O_2). However, there is a difference in that the protective scales are produced in Golgi cisternae,

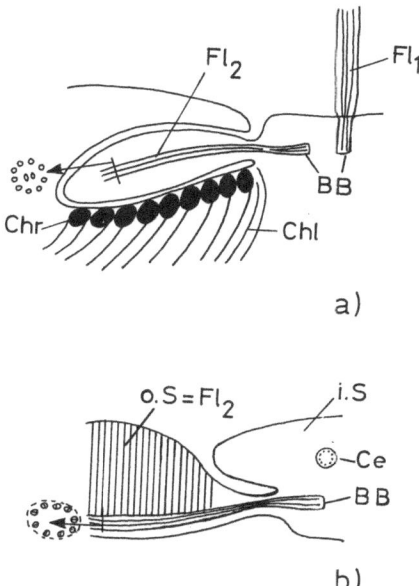

Fig. 25. Homology of the stigma (eye spot) of *Chromulina* and the outer segment of a retinal rod. *BB* basal body (drawn acording to data of ROUILLER and FAURÉ-FREMIET 1958 and FAURÉ-FREMIET 1958). *a*) Stigma. *Fl₁* flagellum, *Fl₂* metamorphized flagellum with 9 + 2 fibrils, *Chl* chloroplast, *Chr* chromatophores of the eye spot. *b*) Retinal rod. *Ce* centriole, *i.S.* inner segment, *o.S.* lamellar outer segment = *Fl₂* with 9 + 0 fibrils.

while the flagellar hairs are a product of the ER system. Hence these two solid secretions also differ in chemical composition according to their origin (polymerous carbohydrates in Golgi, proteins in ER cisternae). In any case scales and mastigonemata have passive functions in protecting the naked plasmalemma or in increasing friction between water and the flagellar surface.

In this special case it seems that the ER membrane of the large tubule-bearing vesicles fuses with the plasmalemma without the usual maturation process in the stack of a dictyosome (see 2.2. and Fig. 7).

4.2.3. De novo Formation of Centrioles

In cells which show centrioles during mitosis, pinwheels can also be found in non-dividing stages. Therefore, it is postulated that the centrioles are transmitted through the cell generations and that their pinwheel structure cannot be formed de novo. However, there seem to be exceptions to this rule.

In the amoebo-flagellate *Naegleria* (FULTON 1971) and *Tetramitus* (OUTKA and KLUSS 1967), the amoeboid cells may transform into flagellates. These produce two flagella each with a clear-cut basal body. On the other hand the amoebae are devoid of centrioles; during mitosis the spindle tubules are generated without this organelle, inside the nucleus (FULTON 1971).

Another case where a de novo formation of centrioles must be considered is the marine protist *Labyrinthula*. During sporulation procentrioles are observed as dense granular aggregates in which cartwheels seem to be assembled spoke by spoke prior to the addition of centriolar microtubules (PERKINS 1970).

Whether in metabiontic cells this faculty of synthesizing centrioles during certain stages of their life cycle has been lost is open to discussion.

4.3. Autoreproduction

After mitosis it is not easy to find the centrioles in differentiated cells which do not develop a flagellum. Nevertheless, it is assumed that they persist through all cell generations and are not formed de novo before they appear during mitosis. This interpretation is based on the spectacular self-reproduction of centrioles in dividing cells and of basal bodies in ciliated cells.

4.3.1. Diplosomes

A good example for showing the continuity of the centriolar apparatus is the sea urchin egg. During the final maturation division of the egg cell, the centriole remains undivided, moves to the other pole of the spindle and disappears (FULTON 1971). In this way the female gamete is left without a centriole. It is restored by the sperm cell which during fertilization brings two centrioles into the zygote (MAZIA *et al.* 1960). Before the first cleavage of the fertilized egg cell the two centrioles duplicate and each mitotic pole is equipped with a diplosome (Fig. 26).

This procedure is repeated during every following mitosis so that an uninterrupted male inheritance is established (MAZIA 1961).

As it is possible to provoke parthenogenetic development of the unfertilized egg cell, the question arises whether sea urchin ontogeny can do without a centriole. Since this is not the case it is not clear whether the egg cell harbors a hidden centriole or whether it is capable of a de novo synthesis if the centriolar inheritance fails.

Studies with so-called *cytasters* which can be produced independently of cell division in nucleate and anucleate egg cells plead for the latter possibility. These microtubular asters are found without centrioles shortly after activation, but with centrioles several hours later (DIRKSEN 1964).

Originally light microscopy led to the assumption that the centriole duplicates by division or sprouting (Fig. 26). However, electron microscopy disclosed a quite different mechanism. When a single centriole evolves into a diplosome, it is observed that at right angles to its axis (see Fig. 24), nine very short microtubular triplets arise in the homogeneous centrosphere and extend in a distal direction, until they reach the length of the tubules in the mother centriole. No contact of the two tubular systems takes place. From the very beginning their definitive diplosomal distance of 50–100 nm is established (FULTON 1971). How the mother centriole organizes the ultrastructure and the growth of its daughter centriole at distance is puzzling, especially as the presence of DNA and messenger RNA is not conclusive.

Since the the centriolar tubules consist of 4.5 nm particulate globular subunits, their extension growth represents an assemblage of macromolecules in a cylindrical configuration. The particles must be preformed in the homo-

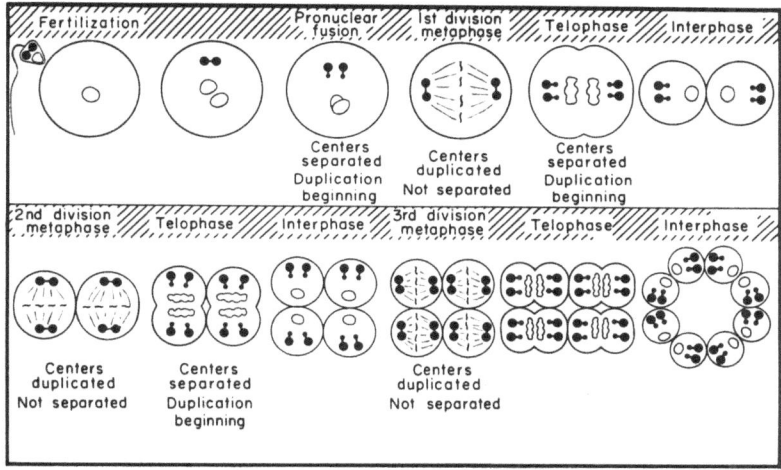

Fig. 26. Sea urchin egg exempt of centrioles is provided with a diplosome by the sperm, whereupon the two centrioles are duplicated (see Fig. 24) synchronically with mitosis [D. MAZIA et al., from J. Biophys. Biochim. Cytol. 7, 1 (1960)].

geneous centrosphere, and since their transition from the solute into the solid state does not need energy, ATP and ATPase would not be necessary for this process.

Bacterial flagella which seem to be homologous to the flagellar fibrils in the eukaryotes can be reconstituted *in vitro* from their particulate subunits (ASAKURA et al. 1964). For this purpose the monomer must be incubated with small pieces of mature flagella as a starter. If such "seeds" are chosen from strains of bacteria with curly flagella and added to the monomer from wild type bacteria which possess straight undulipodia, the reconstituted flagella are curly.

In whatever manner the molecular events of centriolar differentiation may occur, it is evident that every established centriole is an organizer or nucleating center for the differentiation of microtubules. It therefore seems understandable that these organelles may catalyze not only their own duplication, but also the formation of microtubular asters. However, this conclusion is not necessarily right (see 5.3.).

4.3.2. Cilia

In ciliates and ciliated epithelia (nasal passages, trachea, lung, oviduct, of vertebrates) there are large numbers of basal bodies per cell. During the embryonic development of epithelia about 300 cilia can be formed by the two original centrioles. The necessary number of basal bodies is not produced by an exponential series of diplosomes (2, 4, 8, 16 ...) but condensed osmiophilic masses arise in connection with the two inherited centrioles. Around these so-called "condensation forms" procentrioles develop radiating in all directions. During the growth of these centrioles the center of the

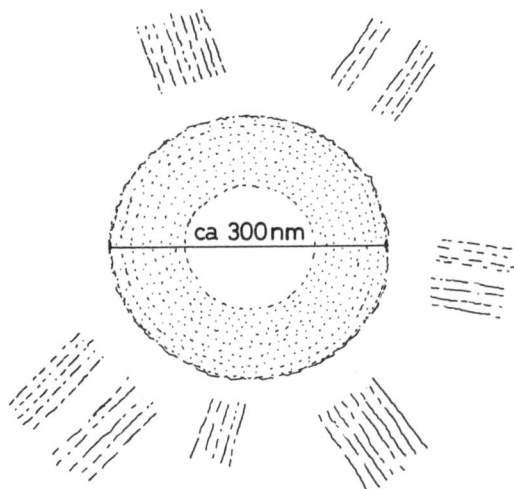

ca 300 nm

Fig. 27. Condensation form surrounded by procentrioles from the oviduct epithelium of a five day old mouse (drawn from electron micrograph by E. R. DIRKSEN, in: CH. FULTON 1971).

condensation form is depleted and may become hollow (Fig. 27). The possibility of a transfer of its osmiophilic material to the procentrioles is discussed by DIRKSEN and CROCKER (1966). The number of condensation forms is such that the necessary amount of centrioles is produced simultaneously. After their formation the procentrioles separate from the condensation form, elongate and move towards the cell surface.

The ontogenetic development of cilia is more complex than that of flagella (see 4.2.1.) because there are intermediate differentiations between the original diplosome and the basal body:

$$\text{diplosome} \rightarrow \text{condensation form} \rightarrow \text{procentriole} \rightarrow \text{centriole} \rightarrow$$
$$\rightarrow \text{basal body} \rightarrow \text{cilium.}$$

How the two original centrioles delegate their organizing capacity to the condensation form is not known. The old problem of classical cytology, whether the simultaneous production of hundreds of basal bodies involves centriolar continuity or a de novo formation, is brought back to the subcellular relation of centriole and condensation form which is the effective organizer.

4.4. Functions

4.4.1. Undulipodia

The function of flagella and cilia as organelles for the locomotion of single cells or the movement of fluids along ciliated epithelia is obvious. How the lash of the flagellum is performed is less clear. As the common name "undulipodium" indicates, the movement is characterized by periodical waves. Originally the beat was described as propeller-like, i.e. as a three-dimensional helical lash. However, it can be shown that only a two-dimensional undular motion takes place and that the helical movement is the result of a rotation of the flagellate (GRAY 1955). The distal tail-end seems involved in causing the rotation (BISHOP 1958). In this way two different motions are super-imposed.

Contractions of the flagellar fibrils are assumed responsible for the oscillation on an even plane. The adoption of this interpretation is based on the chemical nature of the fibrillar protein which is a kind of muscle actin (RENAUD et al. 1968) and the fact that undulipodia contract in a similar way to muscle fibres when ATP is administered to glycerol-extracted flagella, cilia or sperm tails. However, there is a difference in so far as in such experiments the muscular contraction is irreversible while undulipodia display a rhythmic activity (BISHOP 1958).

A beat on an even plane would require an alternating contraction of opposite flagellar sides. However, the symmetry of the flagellum does not favor this view because a plane through the central pair of fibrils divides the outer fibrils unevenly. Therefore, if fibrils 2, 3, 4, and 5 contract individually or jointly in alternation with the opposite fibrils 9, 8, 7, and 6 (Fig. 21), the beat would not be effective in one plane, but in two different directions which diverge by a certain angle.

A further difficulty arises if bacterial flagella are considered. Although they consist of only one fibril they perform similar movements as the eukaryotic flagella with their complicated ultrastructure.

Finally it must be remembered that the contraction of striated muscles is not caused by shortening of individual fibrils but by a two-fibre sliding system (HUXLEY 1957) of the two muscular proteins myosin and actin. Actin forms extremely fine and myosin considerably coarser fibrils. Since certain undulipodia have a set of nine inner secondary filaments which are much finer than the bipartite outer fibrils 1–9 (Figs. 21 b and 23 B, C), the possibility of the muscular "sliding filament" model has been considered. According to SATIR (1968) the doublets 1 and 6 (Fig. 21 a) may slide along each other. However, such speculations cannot explain the motility of the one-stranded prokaryotic flagella.

Since in vivo only the basal body is provided with ATP (see 4.1.1.), the lash of the flagellum seems to be governed by this organelle, and its synonym kinetosome would thus be justified.

4.4.2. Kinetosome

The most important function of the basal bodies is the production of undulipodia. However, besides this generating activity they seem to be involved in producing and propagating the necessary impulses for the induction of flagellar movements. This is especially obvious in ciliates where the beat of neighboring cilia must be coordinated. As this coordination is no longer observed if the undulipodia on the cell surface are rooted too far apart from each other, the seat of coordination cannot be the cortical cytoplasm; it is more likely that it is located in the basal body, which was another reason why in ciliates it was originally termed kinetosome.

In the light microscope it looks as if there were fibrous connections between neighboring basal bodies and one had attributed neural functions to this system. However, in the electron microscope no such structures can be found, so that the nature of the impulse for the ciliar movement and its coordination among the cilia is not yet understood.

For the intraciliar transmission of the impulse from the basal body to the shaft of the flagellum and inside it, the two central fibrils have been considered. However, no proof for this view is available, and again it must be emphasized that prokaryotic flagella perform their beat without such a conductor system.

A second indication of sensorial functions performed by basal bodies is given by their faculty of developing photorecievers such as the eye spot in flagellates (Fig. 25 a) or the outer member of the retinal rods in the eye of vertebrates (Fig. 25 b). Here again it is enigmatic how the light impulses absorbed are transferred from the recievers to the relevant effectors. In flagellated organisms and sprematozoids the basal body is often combined with complicated systems of microtubules organized in strands (*Chlamydomonas*, RINGO 1967 b) or layered structures (*Zamia*, NORSTOG 1967; *Marchantia*, CAROTHERS and KREITNER 1968), so that the kinetosome or blepharoplast is a more complex organelle than the centriole.

4.4.3. Role of the Centriole

Centrioles transmit their basic structure to basal bodies and undulipodia. On the other hand their name derives from quite another function which is also attributed to them.

Since a diplosome forms the centre of the spectacular asters which characterize mitosis in animal cells, it is generally thought that the centrioles are the organizers of cell division. However, this point of view cannot be maintained in general cytology. It must be remembered that only eukaryotes with flagellated male gametes and ciliated epithelia (LENHOSSÉK 1898, FRIEDLÄNDER and WAHRMANN 1965, NAKAO *et al.* 1968) have centrioles. In phyla where the male germ cells have lost their sperm tail such as the siphonogamous plants, centrioles have disappeared as well (LEPPER 1956). In the phylogeny of plants it is striking how from the green flagellates up to the ginkgoales where the gametes are ciliate, centrioles are present, whereas gymnosperms and angiosperms are devoid of them.

In these plants the mitotic spindle is organized by the so-called pole-cap which is a homogenous dense plasmic body comparable to a centrosphere without centrioles. This leads to the conclusion that the centrosphere is the indispensible organelle for the organization of the spindle while the centrioles, if present, are localized in the centrosphere for the following reason:

The centriole is obviously the organizer for the nine stranded basal bodies and the 9 + 2 stranded undulipodia. The microtubules involved in these structures must be quite different from those of the spindle apparatus (see 5.6.). As they can only be produced by the centriolar code this pattern must be transferred from cell to cell. For this reason the duplication of the centrioles is coordinated with that or the chromosomes. In this way mitosis transmits not only chromosomal but also centriolar information from cell to cell. As long as the centriolar model is inherited undulipodia can be produced (flagellates, ciliates, sperm cells, ciliated epithelia) but when the necessity of forming flagella disappears, this organelle is lost as well.

Nematodes are an exception to this rule. As a group they are devoid of cilia and flagella; nevertheless, the mitoses of their cells display classical centrioles within the centrosphere. In this case the centrioles may have undergone a change of function or represent a functionless rudiment.

In conclusion we must distinguish between two different organizers for microtubular systems: the centrosphere for spindle tubules and the centriole for ciliar tubules.

5. Microtubules (MT)

In animal cells and the cortical cytoplasm of plant cells tubular structures have been discovered which are called *microtubules* (SLAUTTERBACK 1963, LEDBETTER and PORTER 1963). It was shown that they display the same morphology as the tubules in the mitotic spindle. Later also the neurotubules, the fibrils in undulipodia (Fig. 21) and even the "flimmer" hairs of hispid flagella in different algal groups (see 4.2.2.) were found to belong to the tubular family.

5.1. Ultrastructure

Cytoplasmic microtubules have a diameter of the order of 20–25 nm, a wall of about 5–8 nm, a bore of less than 10 nm diameter and a length of several µm. These dimensions explain the impossibility of resolving them in the light microscope, so that microtubules were unknown to classical cytology. When so-called "fibers" could be portrayed in the mitotic spindle, these fibrous elements represented bundles of microtubules aggregated by the fixation process.

The fibrillar nature of the spindle was thought to be evidenced by its positive birefringence in polarized light. This double refraction is, however, curious insofar as it is unexpectedly stronger in hydrated living cells (SCHMIDT 1937) than in fixed and dehydrated cells mounted in Canada balsam. This behavior is in opposition to the fact that intrinsic birefringence is a constant quantity which should not depend on the refractive index n of the mounting medium (water n = 1.33, Canada balsam n = 1.54). If such dependence occurs, the birefringence is caused by parallel ultrastructural elements with anisodiametric forms such as rods or lamellae. These elements may even be isotropic, but the difference between their refractive index and that of the medium between the parallel elements causes optical anisotropy. Quantitatively it rises with the the difference of the indices and qualitatively the sign of the birefringence is positive in a system of parallel rodlets and negative in a stack of lamellae. Therefore, this optical anisotropy has been termed form birefringence (Ambronn and Frey 1926).

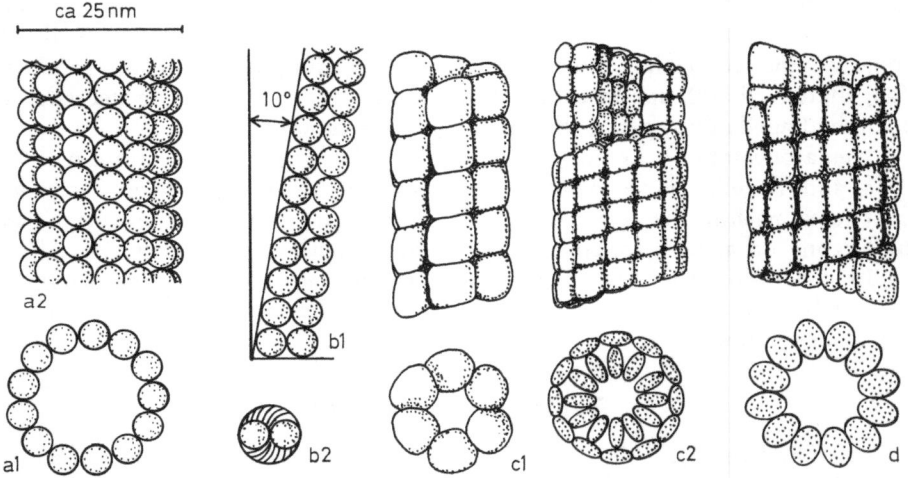

Fig. 28. Ultrastructures of microtubules (MT). *a*) MT in root tip cells of *Juniperus chinensis* with 13 units (Φ 4.5 nm) per turn (Ledbetter and Porter 1964), a_1 cross section, a_2 front view; *b*) Neurofilament = pair of beaded chains from disaggregated neurotubule with a helical angle of 10° (Wisniewski 1968), b_1 front view, b_2 plane view; *c*) MT in yeast cells helical angle 75°–80° (Moor 1967 b), c_1 6 units per turn, c_2 12 units per turn; *d*) P-protein in sieve tubes, helical angle ca. 75° (Parthsarathy and Mühlethaler 1969), 12 units per turn.

Parallel microtubules are an ideal system for producing the effect of form birefringence. The vanishing double refraction of spindles in Canada balsam is easily understood because the protein of the tubules and the balsam have similar refractive indices. The quantitative evaluation in different inhibition media has been studied by Inoué and Sato (1970) who proved that the birefringence of the spindle is a mere form effect. The fact that no appreciable intrinsic anisotropy is found, rules out the presence of fibrillar subunits which always display a strong intrinsic birefringence due to the chain lattice of their cateniform molecules.

The lack of fibrillar properties became understandable when it was found that the wall of the microtubules is not composed of straight chain molecules but of globular macromolecules. Cross-sections of microtubules in the cortical plasm under the plasmalemma of plant cells (Ledbetter and Porter 1964) display 13 particles of 4.5 nm diameter (Fig. 28 *a*).

As a rule the longitudinal files of particles are not arranged parallel to the tubular axis, as sketched in Fig. 28 a_2) but in helical array. In yeast MOOR (1967 b) found a flat coil (Fig. 28 c) with an inclination of $10°–15°$ (helical angle to the tubular axis $75°–80°$). The helix is visible in freeze-etched specimens (Plate IX, Fig. U).

Yeast cells contain three types of microtubules with diameters of 21, 22½ and 25 nm. All three are composed of 8 nm globular units (Fig. 28 c_1). The three different diameters mentioned depend on the number of 5, 6 or 7 units per turn. The unit of 8 nm seems subdivided into 8 subunits of 4 nm

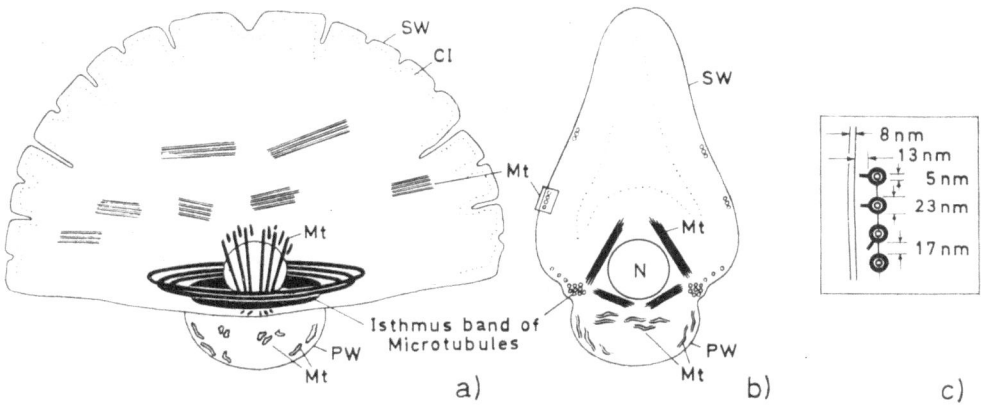

Fig. 29. Drawing of the different microtubular systems in growing cells of *Micrasterias denticulata*. a) Median plane view, b) edge view, c) Diagram of the Mt system in the cortical cytoplasm showing cross-bridges between the Mt and extensions towards the PL. *Mt* Microtubules, *Cl* chloroplast, *PW* primary and *SW* secondary cell wall [courtesy of O. KIERMAYER, from Planta (Berlin) **83**, 223 (1968)].

(Fig. 28 c_2) so that there are 10, 12, or 14 such subunits per turn which numbers compare favorably with the 13 particles found by LEDBETTER and PORTER (1963). Due to the 8 nm particles or two layers of 4 nm particles, the tubular wall is 7.5 nm thick in yeast, whilst in other cases it usually measures 4 nm.

An example is given by the microtubules of the heliozoan axonema (TILNEY 1971). Here it is assumed that 4 nm beads run in 13 files parallel to the tubular axis (cp. Fig. 28 a_2). In contrast to the findings of MOOR (1967 b) no helical structure is observed. However, when the heliozoon *Echinosphaerium* is treated with low temperature, the parallel 22 nm tubules (Fig. 31) disappear and are replaced by 34 nm tubules arranged at random. If these unusually wide microtubules contain the same 4 nm subunits in 13 files as the former tubules, a helical structure is inevitable (TILNEY and PORTER 1967).

An arrangement with a helical angle of $10°$ is found in neurotubules (WISNIEWSKI *et al.* 1968). Here too a 13 stranded model with 4.5 nm beads is presented. As these tubules can disassemble and form 9 nm filaments (Fig. 28 b_1), an association of a standard unit (4.0–4.5 nm beads) may be assumed. In this case paired beads of 4.5 nm particles would make up the 9–10 nm broad filaments (Fig. 28 b_2). The difficulty of visualizing such a model is that 13 is not a multiple of 2.

In bundles of microtubules fibrous or thread-like "spokes" may be observed which interlink the individual tubules. The existence of such bridges has been reported (Tilney 1971) in neuroaxons, the mitotic apparatus, heliozoan axonemata (Fig. 31), phloem tubules (Plate IX, Fig. V) etc. These lateral extensions are especially conspicuous in the desmidian alga *Micrasterias* where in the cortical plasm cross-bridges 17 nm long (Fig. 29) unite the microtubules under the plasmalemma (Kiermayer 1968). The lateral bonds between microtubules may be so pronounced that crystalline bundles result (Brown and Franke 1971).

5.2. Biochemistry

Spindle microtubules and flagellar fibrils can be isolated and analyzed, but this is not the case with plasmic microtubules. Therefore, rash generalizations concerning the chemical composition of different kinds of microtubules should be avoided. Nevertheless, it is generally accepted that they are all composed of proteins which are named *tubulins*.

Mazia and Dan (1952) found that the spindle protein of sea urchin eggs is a macromolecule with the weight 45,000. This molecular weight has been roughly confirmed with 55,000 by Borisy and Taylor (1967). Renaud et al. (1968) succeeded in isolating the outer fibrils of cilia and identified units with the weight of 55,000 ± 5,000. This corresponds to a sedimentation coefficient of 6.0 S and a particle size of 4–5 nm. The amino acid composition resembles that of muscle actin with slight differences of the same order as occur between actins from different species of animals. The content of cysteine is found to be 7.5 sulfhydryl groups per protein unit of 55,000 which may be involved in S-S bonding between particles.

The formation of plasmic and spindle microtubules can be hindered by colchicine. As the spindle protein shows a great affinity to this drug and existing microtubules cannot be disassembled by it, it is assumed that the microtubules are not static but dynamic cell elements which are in equilibrium of constant polar loss and reincorporation of 4.5 nm units. If the disassembled units are blocked by colchicine, no further reintegration will be possible and the microtubules "melt" away.

Plasmic microtubules also disappear at low temperatures (Roth 1967) and high hydrostatic pressures so that three agents are available for intervening experimentally with the microtubular apparatus. Heavy water (D_2O) and high temperatures stabilize these microtubules. On the other hand ciliar microtubules (9 + 2) remain unaffected by the agents mentioned (Tilney and Gibbins 1968).

Not only the diameter of the microtubules is variable (plasmic tubules 25 nm, spindle tubules 15 nm [Harris 1962], yeast tubules 21, 22½, 25 nm, heliozoan tubules 22 and 34 nm, etc.), but also their sensibility to fixation and direct enzymatic digestion in sections, so that Behnke and Forer (1967) distinguish four classes of microtubules with different relative stability in individual cells. It is probable that these variable properties are caused by chemical differences which are not known at the moment.

On the basis of their morphology and conditions of preservation, as many as nine kinds of tubules can be distinguished in the flagellar apparatus of *Naegleria:* "the A, B and C tubules of the basal body, the A and B of the outer fibrils of the shaft, the two central tubules, the spur and the subsurface cytoplasmic tubules" (FULTON 1971, p. 208). A and B tubules differ in amino acid composition (STEPHENS 1970) and the outer doublets of sperm flagella and blastular cilia of sea-urchins seem to hold different tubulins.

5.3. Origin

A pool of 4.5 nm subunits exists in the groundplasm for the formation of microtubules. From this stock the tubules are assembled:

$$\text{4.5 nm subunits} \underset{\longleftarrow}{\overset{\text{GTP}}{\longrightarrow}} \text{microtubules.}$$

As indicated this synthesis does not lead to a static structure but to a dynamic equilibrium. Since an ordered state of the tubular structure is less probable than a random distribution of individual particles, energy is needed for the formation of a tubule. In vitro experiments show that the energy source is not ATP but guanosine triphosphate GTP (STEPHENS 1968). Whether beaded chains or double stranded filaments (WISNIEWSKI *et al.* 1968) which coil up to form the tubular helix are produced first or whether the helical patterns are generated directly from individual subunits is not known. Apparently special conditions govern the helical pitch so that wider or narrower tubules with diameters from 15 to 34 nm (average 25 nm) are built.

In contrast to other cell constituents, such as the endoplasmic reticulum or dictyosomes, the molecular mechanism of tubule synthesis seems to be clear. There must be globular macromolecules of protein with two polar points of junction (FREY-WYSSLING 1955, p. 135) for the formation of the helical chain and additional points for lateral junctions to make up the two-dimensional cylinder sheet. The chemical bondage of these junctions must be weak so that the tubule can easily be disassembled.

The organelles dealt with hitherto have an ontogeny, i.e. they are characterized by an irreversible development leading to complicated structures which age. It is true that their molecules may be remobilized by the cellular turnover. However, this turnover is not a reversible equilibrium process but a unidirectional lysosomal digestion. In contrast to such a monotropic cycle, the growth of a microtubule is the result of a reversible action.

The morphology of the microtubules is similar to that of the capsomeric sheath in rod-shaped viruses. For the lay-out of the capsomeres, virus particles have nucleic acid at their disposal which contains the information for building the proteinic sheath. In the hollow, and therefore "empty" microtubules without nucleic acid, this information seems to be located in so-called *nucleating centers* (TILNEY 1971, p. 242) containing DNA which may be responsible for the microtubular pattern of asters, procentrioles

(Fig. 27), axonemata (Fig. 31), etc. However, the assemblage of TMV protein subunits to an empty tubulus is realized without RNA (STEVENS and LAUFFER 1965).

The nucleating center for the mitotic spindle is the centrosphere with or without collaborating centrioles and for the flagellar fibrils it is a basal body. As centrioles and basal bodies of cilia and flagella are homologous, a homology of mitotic and flagellar tubules has been postulated as well. However, there is no identity between these, because flagellar fibrils do not show the same dynamic assemblage and disassemblage cycle as other microtubules. They are more stable and cannot be impeded by colchicine intervention.

Sometimes cytoplasmic microtubules seem to originate from dense material which stains almost as dark as ribosomes (e.g. in *Acanthamoeba*), but in the cortical plasm of growing cells in higher plants no such nucleating centers have been described as yet.

5.4. P-Protein in Sieve Tubes

In differentiating sieve elements proteinic tubules have been discovered and their globular subunits have been termed P-protein (= phloem protein, CRONSHAW and ESAU 1967, ESAU and CRONSHAW 1967). Although these tubules can be compared to microtubules in every morphological aspect, general cytology has not yet accepted them in the microtubular family. Perhaps this is sound, because the signification of the P-protein might be quite different from that of the tubulins. On the other hand the fact that no notice has been taken of phloem tubules is a sign that the range of modern cytology has become so vast that the vision of an idealized cell with a general ultrastructure is utopian, because phylogeny has diversified plant and animal cells to such a degree, that for functional studies not only histology but also the science of cells must be subdivided into Plant and Animal Cytology.

The tubules of P-protein have diameters of 23 nm in *Nicotiana* and 18–23 nm in *Cucurbita* and a wall thickness of 6–7 nm. Their width corresponds to that of most microtubules. With the radial reinforcement technique (MARKHAM *et al.* 1963) six 7 nm units are visible on the cross-section (PARTHASARATHY and MÜHLETHALER 1969). The units are arranged in a helix and if two subunits per unit are assumed the model of Fig. 28 *d* results.

After differentiation of the sieve elements, the P-protein tubules disassemble. However they are not lost (BOUCK and CRONSHAW 1965) but form striated filaments with a periodicity of 7 nm. Their width is 6 to 12 nm corresponding to single or paired beaded chains of standard units, in full analogy to the filaments derived from neurotubules (Fig. 28 *b*).

Plate IX. Fig. *U*. Helical texture of globular units in microtubules of yeast cells (arrows), freeze-etched, ×100,000 (courtesy of H. MOOR). Fig. *V*. Tubular P-protein in sieve tube element of *Cucurbita maxima*, tubules cohere by spoke-like fibrous extensions, ×80,000 [courtesy of M. V. PARTHASARATHY and K. MÜHLETHALER, from Cytobiology 1, 17 (1969)]. Fig. *W*. Production of flagellar hairs as aligned tubules in the perinuclear space of the nuclear envelope of the chrysophycean flagellate *Olistodiscus luteus*, ×38,000 [courtesy of G. F. LEEDALE, from J. Cell Sci. 6, 701 (1970)].

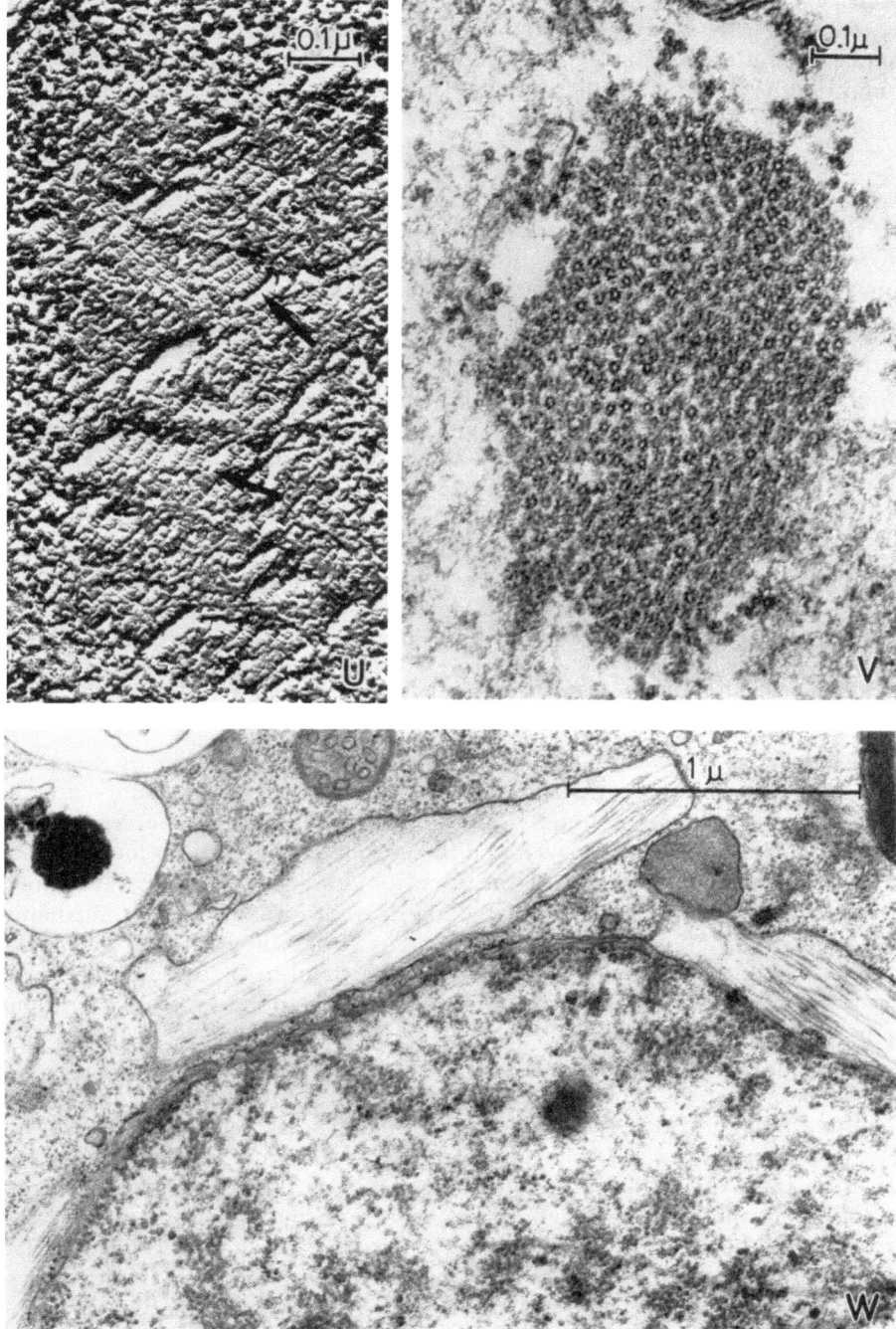

Plate IX.

Before the dispersion into filaments takes place, the tubules of p-protein are parallel in closely packed bundles and interlinked by a kind of fibrous "spokes" (Plate IX, Fig. V) which remind one of the thread-like bridges found in other bundles of microtubules (see 5.1.).

5.5. Functions

Several most diversified functions have been attributed to the microtubules. Of these, their activity in intracellular motility phenomena is prominent. However, with the exception of chromosomal migration, little is known about the mechanism of their intervention in translocation.

5.5.1. Translocation

Since the discovery of the chromosomes in the last century, their migration from the equatorial plane, where they divide, to the two poles of the cell has been a controversial subject. When it was found that the end of the so-called spindle fiber, which turned out to be a bundle of microtubules, joins with the kinetochore of the chromosome, pulling by fiber contraction was suspected. However, as such a contraction of the spindle fibers could not be proved, the idea arose that the growing phragmoplast between the separating chromosomes pushes them apart. As this effect is insufficient and since the phragmoplast, although a distinct entity, has a similar microtubular structure as the spindle body (Esau and Gill 1965, Hepler and Jackson 1968), the possibility of autonomous chromosomal wandering was also considered. None of these three theories is any longer valid.

Inoué and Sato (1967) show that the microtubules which represent the subcellular elements of the spindle fibers indeed attach not only to the kinetochore of the chromosome (Müller 1972) but also to the centrosome. Inspite of this fact, the theory of pulling by fiber contraction could not be verified, because a contraction of the microtubules does not occur. Therefore, McIntosh has devised a model based on the idea that, through mechano-chemical action of intermicrotubular bridges, the tubules slide over one another (Torrey 1971): i.e. the shortening mechanism of muscle fibers was drawn upon where fibrils of actin and myosin glide along each other without changing their length.

However, according to Inoué and Sato (1967), the microtubules shorten visibly without changing their width (Figs. 30 *a, b*). This shortening occurs by disassemblage of the tubular subunits in the polar region. Every disappearing subunit draws its neighbor to the place where it undergoes the transition from the solid to the solute state. The difference to the old theory of pulling is that the microtubules disappear during the migration of the chromosomes, because their subunits return to the pool of dissolved tubular protein particles, which can be re-used for the formation of new microtubules.

Esau and Gill (1965) find that the parallel microtubules of the phragmoplast (Fig. 30) are involved in the movement of vesicles towards the equatorial plane. Both phragmoplast and spindle body appear as discrete entities in which the transportation of vesicles and chromosomes occur in opposite directions. In these movements the microtubules of the phragmoplast seem to have only a guiding function, whereas those of the spindle assume an active part in translocation.

During mitosis the microtubules involved in the transport of vesicles guide Golgi material to the cell plate (PICKETT-HEAPS 1967 a) and to the periplasmic space (ROBARDS and KIDWAI 1969). Other vesicles are seen descending along the polar tubules which do not span from pole to pole, but overlap in the equatorial region of the cell plate (HEPLER and JACKSON 1968).

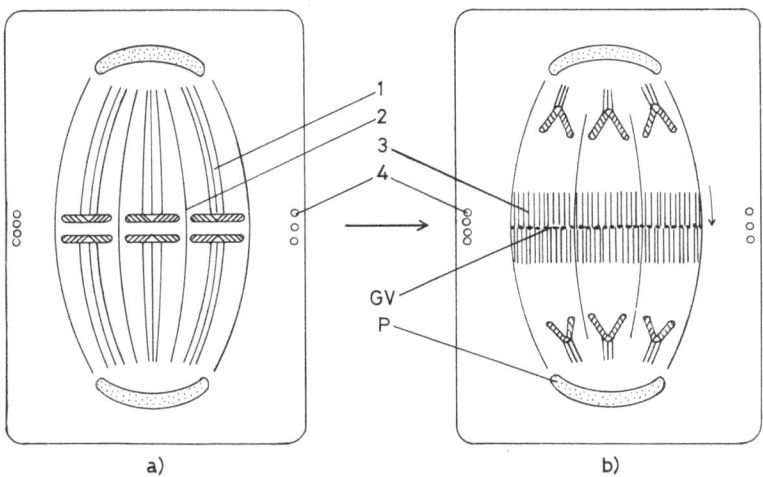

Fig. 30. Microtubular systems involved in mitosis of plant cells. *a*) Metaphase, *b*) anaphase, *P* pole cap, *GV* Golgi vesicles. (1) spindle tubules (INOUÉ and SATO 1967), (2) pole to pole tubules (INOUÉ and SATO 1967); later on the growing cell plate shifts these elements to the periphery of the spindle (GIMÉNEZ-MARTÍN and LOPÉZ-SAÉZ 1963); (3) tubules of the phragmoplast (ESAU and GILL 1965); (4) cortical ring of tubules around the future cell plate (PICKETT-HEAPS and NORTHCOTE 1966 b).

In the pigmented fish *Fundulus* which can adapt its colour to environmental circumstances, the migration of the pigment in the melanophores occurs along microtubules parallel to the direction of movement (BIKLE *et al.* 1966).

It is as if an invisible track in the groundplasm along or between the parallel microtubules were causing the cell elements involved to wander along a defined path.

As the microtubules lack nucleic acid, they cannot act as energy donors for the uphill reactions of directed displacements. So it seems that their guiding function is rather a passive one determining the direction of the movement and the alignment of organelles (TURNER 1968).

5.5.2. Cytoskeleton

The highly anisometric axopodia of suctorians and heliozoans have a central axonema which is composed of hundreds of microtubules (Fig. 31). These are united laterally by two types of bridges: one 7 nm long which links the tubules along a spiral and another of 30 nm length which is responsible for the spacing of the spiral turns (Tilney 1971). From each tubule at least two 7 nm and two 30 nm bonds must lead to its immediate neighbors. As the individual tubule has a 12 or 13-fold symmetry the axonema assumes the profile of a dodecagon. For the construction of such a complicated axonema four different types of macromolecular bonds must radiate from its subunits: First two polar bonds fix the subunits within a beaded chain; then two bonds must stabilize the helical coil of the microtubules; finally the resulting tubules

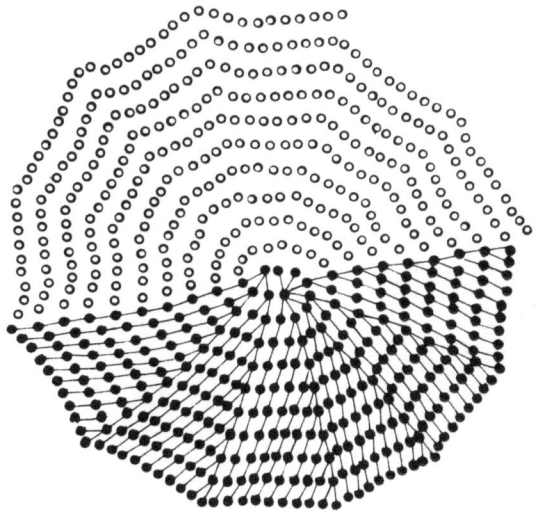

Fig. 31. Microtubules in the axonema of *Echinosphaerium*. In the upper part of the figure the distribution of the MT is seen. In the lower part the direction of the long range bridges of the MT is indicated. In adjacent sectors of the twelve-sided axonema this direction changes by 30°. It is believed that this arrangement of long bridges is related to the substructure of each tubule, which has 12 bonding sites on its surface. ×75,000 (G. Tilney, from: Origin and Continuity of Cell Organelles, p. 250, Berlin-Heidelberg-New York: Springer 1971).

must be assembled by 7 nm and 30 nm long-range forces into the 12-sided spiral arrangement of the rod-like axonema (Fig. 31). The long-range bridges are thread-like and can be made visible in the high resolution electron microscope.

Such a solid rod certainly has skeletal functions, but in analogy to neuro-tubules it may also have something to do with stimulus conduction.

Another example where the microtubular system has been regarded as a cytoskeleton are the asymmetrical gastrulae of sea urchins. If they are treated with colchicine the gastrulae spherulate, which effect can be hindered by D_2O which stabilizes the tubular system. This seems to indicate that the micro-tubules have a form giving function. However, later it was found that they are only influential during the development of the gastrulae and that they do little for the maintenance of the cell shape in differentiated tissues (Tilney and Gibbins 1969). So their function seems rather to be connected with morphogenesis (see 5.5.4.).

PORTER and TILNEY (1965) supported the view "that microtubules serve as elastic cytoskeletal elements and that energy released on their surfaces is translated, by an as yet unknown mechanism, into cytoplasmic movements". Here we encounter the theories concerning microtubular activity in cytomotility.

5.5.3. Cytomotility

Since the fibrils in the flagella are considered as prototypes of microtubules, intracellular motility was thought to be a main field of microtubular activity. This view is supported by the nature of their protein tubulin which, as already mentioned, resembles that of muscle actin (see 5.2.). So the assumption is that microtubules might act in the cell in a similar way as muscle fibers on the histological level. However, such a comparison has its drawbacks, because contractility is unknown in plasmic microtubules, and *in vitro* contraction of glycerol extracted flagella with ATP does not concern individual microtubular fibrils but the undulipodian apparatus as a whole (see 4.4.1.).

All the same, not only flagellar motility but also plasmic streaming (cyclosis, rotation, circulation, etc.) has been attributed to microtubular activity. PORTER's earlier research on plant cells which seemed to indicate that microtubules are essential to streaming has been extended to the movement of cytoplasmic granules in the chromatophores of fish (see 5.5.1.).

Also other motile systems are rich in microtubules. In spermatozoids of plants a multilayered structure is found around the nucleus of the sperm (BELL *et al.* 1971). In pteridophytes this lamellar body is tripartite, the outer layer consisting of a long band of parallel microtubules. In sperms of *Equisetum* the basal bodies of the multiflagellate locomotion apparatus are arranged along this microtubular layer (DUCKETT and BELL 1969).

5.5.3.1. *Filaments*

In strands of the plasmodium of slime moulds (*Physarum*), where a liquid plasma sol flows periodically to and fro with high velocity in a pipe of plasma gel, the presence of many microtubules has favored the hypothesis of their function in plasmic motility. However, in the ectoplasm of amoeba which controls the pseudopodial movements much finer plasmic filaments with 8 nm diameter (against 24 nm of the microtubules) seem rather to be responsible for the motility of this organism (WOHLFARTH-BOTTERMANN 1967).

ALLISON (1971) finds that the glycerol extracted cortical cytoplasm of many cells is rich in such filaments. At the place where endocytosis takes place they form a circular system around the invagination of the plasmalemma and seem to squeeze its neck and help to form the PL-bound endocytotic vesicles (Fig. 4 *b, c*).

Whether such filaments are related to disassembled microtubules, as in the case of neurotubules (Fig. 28 *b*), is not known. Another problem is whether the plasmic filaments cause movements by their own contraction or by a sliding mechanism.

5.5.4. Morphogenesis

The ectoplasm controls the shape of a cell. Therefore, not only the passive function of a cytoskeleton but also its active participation in morphogenesis has been attributed to the microtubules which it contains. The sea urchin gastrula which loses its typical asymmetrical shape when the microtubular apparatus is destroyed by colchicine, low temperature or high pressure, has already been referred to (see 5.5.2.). As this effect is only operative during early stages of development (Tilney and Gibbins 1969), such observations plead strongly for a morphogenetic influence exerted by the microtubules, although the agents which force the tubular particles to disassemble may also interplay with other, as yet unknown, cell structures involved in morphogenesis.

In the cortical plasm of growing plant cells, systems of parallel microtubules are found directly under the plasmalemma (Ledbetter and Porter 1963). Also the place where during mitosis the cell plate will meet the longitudinal wall of the dividing cell (Fig. 30) is marked by a ring of microtubules (Pickett-Heaps and Northcote 1966 b, Cronshaw and Esau 1968). In general their direction coincides with that of the cellulose elementary fibrils synthesized in the pericellular matrix of the cell wall outside the plasmalemma. Therefore, it was thought that the microtubules represented a preformed pattern for the morphogenesis of the reinforcing and shape giving fibrillar elements in the amorphous gel of the cell wall matrix. This view is supported by the intracellular presence of microtubules in a circular array (Fig. 17) where a ring-shaped reinforcement of the secondary wall is produced in water conducting xylem elements (Pickett-Heaps and Northcote 1966 a, Pickett-Heaps 1967 b).

How the microtubules can act at a distance across the plasmalemma is puzzling. Since other authors have found that the cellulose elementary fibrils are orientated not parallel but perpendicular to the cortical microtubules (Green 1962), or in cross-textured secondary walls alternatively parallel and perpendicular to them, the theory of a direct influence of the microtubules on the wall texture has been criticized (Chafe and Wardrop 1970).

A new avenue has been opened by the discovery that during cell elongation a large quantity of microtubules may appear in the pericellular space between plasmalemma and cell wall (Heyn 1971). Microtubules outside the plasmalemma were already observed in 1965 (Cronshaw and Bouck 1965) so that in cell wall synthesis, they may be endowed with extracellular functions, e.g. enzyme donation.

5.6. Homonymy

The review of possible microtubular functions embraces a wide spectrum of physiological activities with obvious contradictions, since an elastic cytoskeleton is scarcely compatible with motility. On the level of histology, evolution has developed two different protein classes to meet the requirements of these two functions, namely collagens and actomyosins. Therefore, it is unlikely that the same class of proteins, namely tubulins, can fulfil both skeletal stability and motile activity.

The catalogue of possible tubular functions further comprises neural stimulation, morphogenesis, enzymatic actions etc. If the P-protein tubules (phloem protein) are included (see 5.4.) an additional function would be protein accumulation, because the P-tubules store the protein units for the filaments which, after maturation of the sieve tubes, become functional in phloem activity.

In the ER cisternae of the cells in the connective tissue of snails, ephemeric microtubules (25 nm $\emptyset$, 12 units of 6 nm $\emptyset$, 3 nm periodicity) which are thought to represent an intermediate stage of protein synthesis temporarily assume a tubular pattern (STANG-VOSS and STAUBESAND 1971).

All these interpretations together with the wide span of tubular size (from 18 nm in spindle tubules to 34 nm in low temperature induced heliozoan tubules) and the helical arrangements of the subunits (helical angles from 10° to 80°) makes it unlikely that "the microtubule" is endowed with the task of a special organelle. In mitosis four sets of microtubules (Fig. 30) with different functions can be observed (JOHNSON and PORTER 1968).

The term "microtubuli", therefore, is rather a morphological concept (FREY-WYSSLING 1972). Whenever globular protein molecules are assembled, they form beaded chains and the handling of such unwieldy threads leads necessarily to coils and the formation of helically textured tubules. As this principle is also realized in the capsomeric sheath of rod-shaped viruses and shown by the disassemblage of neurotubules and P-protein tubules into functional filaments, I claim that the microtubulus is primarily only a general form for a two-dimensional arrangement of protein macromolecules, similar to one-dimensional beaded chains or three-dimensional crystal lattices of globular units. Compared with a membrane, the tubule is a more condensed arrangement and can be conceived as a two-dimensional lattice. Evidently such a hollow cylinder is more flexible than a three-dimensional lattice and its macromolecules are accessible from both the outside and the inside which facilitates its rapid disassemblage. As a consequence, and in contrast to real organelles, the tubuli easily disintegrate into their building units:

$$\begin{array}{c} \text{macromolecular units or subunits} \\ \uparrow\downarrow \\ \text{microtubules} \\ \downarrow \\ \text{filaments} \end{array}$$

Since, as already stated, their disassemblage into particles is *reversible,* no ontogeny can be attributed to them because ontogeny is a uni-directional non-reversible development of a biological system.

It is true that membranes can also disassemble into their molecular elements; but for their reassembling a preformed starter is necessary

$$\text{membrane} \rightleftarrows \text{molecular elements} \\ \uparrow \\ \text{starter}$$

Thus, for the formation of an organelle a more complicated organization is necessary than just a reversible assemblage of individual particles. The discussion on the biological status of ribosomes and polysomes (see 3.5.1.) has already shown that different types of macromolecules must work together to perform energy consuming functions. As this prerequisite is not realized in individual microtubuli, they are, in spite of their considerable size, not organelles.

Therefore, it must be postulated that not the microtubules themselves but their association with the intertubular groundplasm represent an organelle, in the same way as not individual Golgi cisternae but only their dictyosomal stacks together with the intercisternal plasm have the status of functional organelles.

Such microtubular organelles are the mitotic spindle, the phragmoplast, the heliozoan axonema, the organization of flagellar 9 + 2 strands etc. As to the loose grids of parallel tubules under the plasmalemma in the cortical ectoplasm, or the clusters of microtubules in the perinuclear space (Plate IX, Fig. W) or even in the pericellular space between plasmalemma and cell wall (HEYN 1970), we must first find out the nature of their cooperation with energy donors (ATP, GTP or others), before we can speak of an organelle. When their requirement and source of energy is known, the present ambiguity of their possible activities will certainly be replaced by the knowledge of their real function.

Meanwhile the microtubules which are not organized in recognizable organelles must be considered as a family of mere morphological similarity and the term "microtubule" has no better informative value than the name "unit membrane" in membranology.

Since individual microtubules have no ontogenetic development, different types of them cannot be considered as homologous, and as their assumed physiological activities cover a vast variety of hypothetical functions, their various representatives are hardly analogous. Their similar shape only, has led to a common name which says nothing about their probably diversified significance. It does not describe a group of ontogenetically or functionally related cell elements but an ultrastructural principle of gathering protein macromolecules in a subcellular cylinder.

Therefore, for the time being, the vast family of microtubules cannot be characterized by the principle of homology, nor by that of analogy, but represents a heterogeneous group brought together by mere *homonymy*.

6. Retrospect

The problems concerning "Macromolecules in Cell Structure" (FREY-WYSSLING 1957) have evolved in the past 15 years to questions on the organization of the "Macromolecules in Organelles". At the same time the highly dynamic nature of the organelles during growth and differentiation of the cell became evident. As a consequence ultrastructure research can no longer concentrate on mere morphology but must consider structural changes during ontogeny and functional activity.

On the other hand developmental and dynamic studies are only possible on the basis of established morphological patterns which, through their labile and changeable structure, can assume different forms or evolve into more complicated arrangements. As to the patterns mentioned, three different types can be distinguished: namely *linear, two-dimensional* and *three-dimensional* arrangements of macromolecules or elementary cell constituents.

The one-dimensional linear type is realized in bundles of chain molecules (cellulose, chitin, silk fibroin, collagen, myosin, actin), beaded chains of globular macromolecules and strands of nucleic acids. The resulting filaments, elementary fibrils, microfibrils and fibers play their roles in the functions of heredity, skeletal reinforcement and motility. It may be mentioned that the first attack on the problems of cellular ultrastructure was concerned with such linear systems when a satisfactory insight into their molecular biology had been gained (FREY-WYSSLING 1938).

The two-dimensional patterns and their functions are more difficult to analyze and explain, because X-ray diffraction has not yet yielded such clear results as in the case of one-dimensional systems. Microtubules and biomembranes (cytomembranes) belong to these systems. It is a field in which research is intensively engaged at the moment and in which clarifying results are expected in the near future (MÜHLETHALER 1972).

Microtubules are sheets of proteinic macromolecules which are arranged in a cylindrical plane. They are accessible from both the outside and the inside of the hollow cylinder. The tubules may reversibly disintegrate into their macromolecular units and subunits or into filaments of beaded chains and reintegrate into tubules. As a rule the beaded chains form a helix in the tubular wall; only rarely do they run parallel to the tubular axis.

The function of systems of such tubules are manifold (see 5.5.). The simplest is the storage of enzyme molecules, reserve protein or other molecular units which form a pool in the groundplasm. Other tasks are much more complicated and so different from each other that there are no analogous functions. This is especially true for the tubules of the cortical plasm, the microtubular fibrils of the undulipodia and the tubules of the mitotic spindle. For the shortening of the latter in connection with the translocation of the chromosomes a disassemblage of the tubular subunits is more likely (see 5.5.1.) than a mutual sliding as in the contraction of striated muscle fibres, because the sliding linear elements in muscle (myosin and actin) are not tubules but solid microfibrils.

Individual microtubules are not organelles. They can only function in cooperation with other tubules and the intertubular, energy-supplying groundplasm. Therefore, a tubule has no special physiological status. It is just a morphological term such as "granule", "globule", "vesicle" or "unit membrane" (FREY-WYSSLING 1972).

In the same way as an organ must have free energy (e.g. by respiration) and coordinative stimulation (e.g. by nerves) for successful operations, a subcellular structure can only work as an organelle, if it is equipped with an energy donor (e.g. ATP) and provided with the necessary information (e.g. by messenger RNA) for the performance of cooperative activity. As a consequence, for the definition and a full understanding of an organelle, three requirements must be met with:

1. a labile *structure* of proteins, nucleic acids and/or carbohydrates as a morphological base,

2. an *energy donor* for its dynamic actions, and

3. a system of *information* for the regulation of its activity within the cellular metabolism.

The concept of the unit membrane as an equilibrated static three-lamellated film has turned out to be too simple for the understanding of the functional activities of biomembranes. The cytomembranes which lead to the compartmentation of the cell are subject to ontogeny, i.e. they develop, age and display a turnover. They function with the aid of energy sources which balance their labile non-equilibrated structure. For their activities in translocation of ions and molecules, local anabolism, resorption and elimination phenomena, they contain enzyme complexes of protein which give the membrane complicated specific features. The specificity can be proved by immunological reactions (GITZELMANN *et al.* 1970).

Ontogenetically there seem to be two classes of such membranes: PL and ER membranes (Table 4). As a rule no fusion takes place between these. The membranes of the plasmalemma and of fully differentiated Golgi cisternae are thicker (around 10 nm) than those of the endoplasmic reticulum (around 7.5 nm). The dictyosomes can transform ER membranes into PL membranes during the shift of their cisternae from the proximal to the distal pole (Fig. 7). This capacity hallmarks the Golgi apparatus not only as an organelle in the service of elimination but also as a producer of intracellular plasmalemma which is incorporated into the cell membrane and permits its rapid growth in area.

This statement does not seem valid for lower plants such as the ascomycetes, where the transformation ER → PL occurs without a Golgi intermediate (Figs. 15 *a–c*). Such facts raise interesting ontogenetic and even phylogenetic problems for comparative organellography.

While the intense study of two-dimensional structures has led to the new cytological branch "membranology", research on the three-dimensional ultrastructures will be a task for the more distant future.

It is true that the spatial ultrastructures of static cell constituents such as the space lattice of protein crystals, the basal body of the undulipodia (Fig. 24) or the prolamellar body with its cubic symmetry (GUNNING 1965) in the chloroplast, can be thoroughly described; but what the labile arrangement, the interrelations and the dynamics of nucleic acids, enzymes and "starters" in the groundplasm are, will be a main concern of tomorrow's Molecular Biology.

Bibliography

AFZELIUS, B. A., 1955: The ultrastructure of the nuclear membrane of the sea urchin oocyte as studied with the electron microscope. Exp. Cell Res. **8**, 147.

ALBERSHEIM, P., 1965: A cytoplasmic component stained by hydroxylamine and iron. Protoplasma **60**, 131.

ALLEN, R. D., 1968: A reinvestigation of cross sections of cilia. J. Cell Biol. **37**, 825.

ALLISON, A. C., 1971: Plasm filaments. Oral communication at the symposium "Recent physical studies on the structure of biomembranes", Titisee, West-Germany.

AMBRONN, H., and A. FREY, 1926: Das Polarisationsmikroskop. Leipzig: Akadem. Verlagsges.

ANTROPOVA, E. N., and Y. F. BOGDANOV, 1971: Cytophotometry of DNA and histone in meiosis of *Pyrrhocoris apterus*. Exp. Cell Res. **60**, 40.

ASAKURA, S., G. EGUCHI, and T. INO, 1964: Reconstitution of bacterial flagella *in vitro*. J. Mol. Biol. **10**, 42.

BAUER, H., 1967: Ultrastruktur und Zellwandbildung von *Acanthamoeba* sp. Dr. thesis ETH Zürich; Vierteljahrsschr. Naturf. Ges. Zürich **112**, 173.

BEHNKE, O., and A. FORER, 1967: Evidence for four classes of microtubules in individual cells. J. Cell Sci. **2**, 169.

BELL, P. R., 1970: Are plastids autonomous? Symp. Soc. exp. Biol. **24**, 109.

— J. G. DUCKETT, and D. MYLES, 1971: The occurrence of a multilayered structure in the motile spermatozoids of *Pteridium aquilinum*. J. Ultrastruct. Res. **34**, 181.

— A. FREY-WYSSLING, and K. MÜHLETHALER, 1966: Evidence of the discontinuity of plastids in the sexual reproduction of a plant. J. Ultrastruct. Res. **15**, 108.

— and K. MÜHLETHALER, 1962: The fine structure of the cells taking part in oogenesis in *Pteridium aquilinum* (L.) Kuhn. J. Ultrastruct. Res. **7**, 452.

— — 1964: Evidence for the presence of deoxiribonucleic acid in the organelles of the egg cells of *Pteridium aquilinum*. J. Mol. Biol. **8**, 853.

BEN-HAYYIM, G., and I. OHAD, 1965: Synthesis of cellulose by *Acetobacter xylinum*. J. Cell Biol. **25**, No. 2, Part 2, 191.

BERNHARD, W., and E. DE HARVEN, 1960: L'ultrastructure du centriole et d'autres éléments de l'appareil achromatique. Proc. 4th Intern. Congr. Electron Micr. Berlin 1958. Verhdlgn. **2**, 217. Berlin-Göttingen-Heidelberg: Springer.

BIKLE, D., L. G. TILNEY, and K. R. PORTER, 1966: Microtubules and pigment migration in the melanophores of *Fundulus heteroclitus* L. Protoplasma **61**, 322.

BISALPUTRA, TH., and A. A. BISALPUTRA, 1967: The occurrence of DNA fibrils in chloroplasts of *Laurencia spectabilis*. J. Ultrastruct. Res. **17**, 14.

BISHOP, D. W., 1958: Motility of the sperm flagellum. Nature **182**, 1638.

BLACK, S. H., and C. GORMAN, 1971: Nuclear segregation and envelopment during ascosporogenesis in *Hansenula wingei*. Arch. Mikrobiol. **79**, 231.

BORISY, G. G., and E. W. TAYLOR, 1967: The mechanism of action of colchicine. Colchicine binding to sea urchin eggs and the mitotic apparatus. J. Cell Biol. **34**, 535.

BOUCK, G. B., 1962: Chromatophore development, pits, and other fine structures in the red alga, *Lomentaria baileyana*. J. Cell Biol. **12**, 553.

— 1969: Extracellular microtubules. J. Cell Biol. **40**, 446.

— and J. CRONSHAW, 1965: The fine structure of differentiating sieve tube elements. J. Cell Biol. **25**, 79.

BRACKER, C. E., 1967: The ultrastructure and development of sporangia in *Gilbertella persicaria*. Mycologia **60**, 1016.

BRANDT, P. W., and G. D. PAPPAS, 1960: An electron microscopic study of pinocytosis in *Amoeba*. J. Biophys. Biochem. Cytol. **8**, 675.

BRANTON, D., 1966: Fracture faces of frozen membranes. Proc. nat. Acad. Sci. **55**, 1048.

BREIDENBACH, R. W., A. KAHN, and H. BEEVERS, 1968: Characterization of glyoxisomes from castor bean endosperm. Plant Physiol. **43**, 705.

BROWN, R. M., JR., 1969: Observations on the relationship of the Golgi apparatus to wall formation in the marine chrysophycean alga, *Pleurochrysis scherffelii* Pringsheim. J. Cell Biol. **41**, 109.
— and W. W. FRANKE, 1971: A microtubular crystal associated with the Golgi field of *Pleurochrysis scherffelii*. Planta (Berlin) **96**, 354.
— — H. KLEINIG, H. FALK, and P. SITTE, 1970: Scale formation in chrysophycean algae. I. Cellulosic and noncellulosic wall components made by the Golgi apparatus. J. Cell Biol. **45**, 246.
BUVAT, R., 1962: L'origine et le développement des vacuoles des cellules végétales. 5th Intern. Congr. Electron Microscopy Philadelphia 1962, Vol. 2, W-1. New York: Academic Press.

CAROTHERS, Z. B., and G. L. KREITNER, 1968: Studies of spermatogenesis in the Hepaticae II. Blepharoplast structure in the spermatid of *Marchantia*. J. Cell Biol. **36**, 603.
CARROLL, G. C., 1967: The ultrastructure of ascospore delimitation in *Saccobolus kerverni*. J. Cell Biol. **33**, 218.
CECCHI FIORDI, A., and E. MAUGINI, 1972: Relationships between the endoplasmic reticulum and plastids of the ovule of *Ginkyo biloba*. Caryologia (in press).
CHAFE, S. C., and A. B. WARDROP, 1970: Microfibril orientation in plant cell walls. Planta (Berlin) **92**, 13.
CLAUDE, A., 1946: Fractionation of mammalian liver cells by differential centrifugation. J. Exp. Med. **84**, 51, 253.
CLOWES, F. A. L., and B. E. JUNIPER, 1968: Plant Cells. Botan. Monographs Vol. 8. Oxford-Edinburgh: Blackwell.
COHEN, S. S., 1970: Are/were mitochondria and chloroplasts microorganisms? Amer. Scientist **58**, 281.
COLVIN, J. R., 1964: The biosynthesis of cellulose. In: The formation of wood in forest trees (ZIMMERMANN, M. H., ed.), p. 189. New York: Academic Press.
CORTAT, M., 1971: Localisation intracellulaire et fonction des β-1,3 glucanases lors du bourgeonnement chez *Saccharomyces cerevisiae*. Dr. thesis ETH Zürich.
CRONSHAW, J., and G. B. BOUCK, 1965: The fine structure of differentiating xylem elements. J. Cell Biol. **24**, 415.
— and K. ESAU, 1967: Tubular and fibrillar components of mature and differentiating sieve elements. J. Cell Biol. **34**, 801.
— — 1968: Cell division in leaves of *Nicotiana*. Protoplasma **65**, 1.

DANIELLI, J. F., and E. N. HARVEY, 1935: The tension at the surface of mackerel egg oil, with remarks on the nature of the cell surface. J. cell. comp. Physiol. **5**, 483.
DAVSON, H., and J. F. DANIELLI, 1943: The permeability of natural membranes. Cambridge: University Press.
DE DUVE, C., 1959: Lysosomes, a new group of cytoplasmic particles. In: Subcellular particles (HAYASHI, T., ed.), p. 128. New York: Ronald Press.
— 1963: The lysosome. Sci. American **208**, 2.
— and P. BAUDHUIN, 1966: Peroxisomes (microbodies and related particles). Physiol. Rev. **46**, 323.
— and R. WATTIAUX, 1966: Functions of lysosomes. Ann. Rev. Physiol. **28**, 435.
DE VRIES, H., 1885: Plasmolytische Studien über die Wand der Vakuolen. Jb. wiss. Bot. **16**, 465.
DIERS, L., 1965: Elektronenmikroskopische Untersuchungen über die Eizellbildung und Eizellreifung des Lebermooses *Sphaerocarpus donnellii* Aust. Z. Naturforsch. **20 b**, 795.
DIRKSEN, E. R., 1964: The isolation and characterization of asters from artificially activated sea urchin eggs. Exp. Cell Res. **36**, 256.
— and T. T. CROCKER, 1966: Centriol replication in differentiating ciliated cells of mammalian respiratory epithelium. J. Microscopie **5**, 629.
DRAWERT, H., 1953: Vitale Fluorochromierung der Mikrosomen mit Janusgrün, Nilblausulfat und Berberinsulfat. Ber. dtsch. bot. Ges. **66**, 134.

DREWS, G., M. BIEDERMANN, and J. OELZE, 1969: Investigation of the thylakoid morphogenesis in *Rhodospirillum rubrum*. In: Progress in photosynthetic research (METZNER, H., ed.), Vol. I, p. 204. Tübingen.

— and P. GIESBRECHT, 1963: Zur Morphogenese der Bakterien-„Chromatophoren" (-Thylakoide) und zur Synthese des Bakterienchlorophylls bei *Rhodopseudomonas spheroides* und *Rhodospirillum rubrum*. Bakteriol. u. Hygiene **190**, 508.

DUCKETT, J. G., and P. R. BELL, 1969: The occurrence of a multilayered structure in the sperm of a petridophyte. Planta (Berlin) **89**, 203.

ESAU, K., and J. CRONSHAW, 1967: Tubular components in cells of healthy and tobacco mosaic virus-infected *Nicotiana*. Virology **33**, 26.

— and R. H. GILL, 1965: Observations on cytokinesis. Planta (Berlin) **67**, 168.

FAHN, A., and T. RACHMILEVITZ, 1970: Ultrastructure and nectar secretion in *Lonicera japonica*. New Res. Plant Anat. 1, 51, Suppl. 1 to Bot. J. Linnean Soc. Vol. 63.

FAURÉ-FREMIET, E., 1958: The origin of the metazoa and the stigma of the phytoflagellates. Quart. J. Microsc. Sci. **99**, 123.

FAWCETT, D., 1961: Cilia and flagella. In: The cell (BRACHET, J., and A. E. MIRSKY, eds.), Vol. II, p. 217. New York-London: Academic Press.

— 1965: Surface specialization of absorbing cells. J. Histochem. Cytochem. **13**, 75.

— and K. R. PORTER, 1952: Number of fibrils in sperm tails. Anat. Record **113**, 539, Abstract 33.

— and I. SUSUMO, 1958: Observations on the cytoplasmic membranes of testicular cells, examined by phase contrast and electron microscopy. J. Biophys. Biochem. Cytol. **4**, 135.

FELDHERR, C. M., 1965: The effect of the electron-opaque pore material on exchanges through the nuclear annuli. J. Cell Biol. **25**, 43.

FINEAN, J. B., 1953: Further observations on the structure of myelin. Exp. Cell Res. **5**, 202.

FINERAN, B. A., 1970: Organization of the tonoplast in frozen-etched root tips. J. Ultrastruct. Res. **33**, 574.

— 1972: Association between ER and vacuoles in frozen-etched root tips. J. Ultrastruct. Res. (in press).

FLUCK, D. J., A. F. HENSON, and D. CHAPMAN, 1969: The structure of dilute lecithin-water systems revealed by freeze-etching and electron microscopy. J. Ultrastruct. Res. **29**, 416.

FRANKE, W. W., 1970 a: On the universality of nuclear pore complex structure. Z. Zellforsch. **105**, 405.

— 1970 b: Nuclear pore flow rate. Naturw. **57**, 44.

— and J. KARTENBECK, 1969: Structure of nuclear membranes isolated from brain cells. Experientia **25**, 396.

— and U. SCHEER, 1970: The ultrastructure of the nuclear envelope of amphibian oocytes: a reinvestigation. J. Ultrastruct. Res. **30**, 288.

— — 1972: Structural details of dictyosomal pores. J. Ultrastruct. Res. (in press).

FRANZ, G., and H. MEIER, 1969: Einige Probleme der Zellulosebiosynthese. Verhdlgn. Schweiz. Naturf. Ges., p. 159.

FREY-WYSSLING, A., 1935: Die Stoffausscheidung der höheren Pflanzen. Berlin: Springer.

— 1938: Submikroskopische Morphologie des Protoplasmas und seiner Derivate. Berlin: Borntraeger.

— 1942: Zur Physiologie der pflanzlichen Glukoside. Naturw. 3, 500.

— 1948: The growth in surface of the plant cell wall. Growth Symp. **12**, 151.

— 1955: Die submikroskopische Struktur des Cytoplasmas. Protoplasmatologia Vol. II A2. Wien: Springer.

— 1957: Macromolecules in cell structure. Prather Lectures 1956. Cambridge, Mass.: Harvard University Press.

— 1962: Interpretation of the ultrastructure in growing plant cell walls. In: Interpretation of ultrastructures. Symposia Internat. Soc. Cell Biol. 1, 307.

— 1963: Active intake of molecules and ions by pinocytosis. Lecture 6, Permeability Conf. Wageningen. Zwolle, Netherl.: Publ. Willink.

FREY-WYSSLING, A., 1967: Gymnoplasts instead of "protoplasts". Nature **216**, 516 (only).
— 1969: The ultrastructure and biogenesis of native cellulose. Fortschr. Chemie org. Natur-
stoffe, Vol. 27, p. 1. Wien-New York: Springer.
— 1972: Über Microtubuli. Rev. Suisse de Zoologie **79**, 29.
— E. GRIESHABER, and K. MÜHLETHALER, 1963: Origin of spherosomes in plant cells.
J. Ultrastruct. Res. **8**, 506.
— J. F. LÓPEZ-SÁEZ, and K. MÜHLETHALER, 1964: Formation and development of the cell
plate. J. Ultrastruct. Res. **10**, 422.
— and K. MÜHLETHALER, 1965: Ultrastructural plant cytology. Amsterdam: Elsevier.
— and E. STEINMANN, 1953: Ergebnisse der Feinbauanalyse der Chloroplasten. Vierteljahrsschr. Naturf. Ges. Zürich **98**, 20.
FRIEDLÄNDER, M., and J. WAHRMAN, 1965: The independence of spermatid differentiation
from the meiotic divisions. Exp. Cell Res. **38**, 680.
FULTON, C., 1971: Centrioles. In: Origin and continuity of cell organelles (REINERT, J.,
and H. URSPRUNG, eds.), p. 170. Berlin-Heidelberg-New York: Springer.

GIBBONS, I. R., 1967: The organization of cilia and flagella. In: Molecular organization
and biological function (ALLEN, J. M., ed.), pp. 211—237. New York: Harper & Row.
— and A. V. GRIMSTONE, 1960: On flagellar structure in certain flagellates. J. Biophys.
Biochem. Cytol. **7**, 697.
GIBBS, S. P., 1962: Nuclear envelope—chloroplast relationships in algae. J. Cell Biol. **14**,
433.
GIESBRECHT, P., and G. DREWS, 1966: Über die Organisation und die makromolekulare
Architektur der Thylakoide „lebender" Bakterien. Arch. Mikrobiol. **54**, 297.
GIMÉNEZ-MARTÍN, G., and J. F. LÓPEZ-SÁEZ, 1963: Achromatic spindle in higher plants.
ΦYTON **20** (2), 183.
GIRBARDT, M., 1969: Die Ultrastruktur der Apikalregion von Pilzhyphen. Protoplasma **67**,
413.
GITZELMANN, R., TH. BÄCHI, H. BINZ, J. LINDENMANN, and G. SEMENZA, 1970: Localization
of rabbit intestinal sucrase with ferritin-antibody conjugates. Biochim. Biophys. Acta **196**,
20.
GOEBEL, K., 1913/1923: Organographie der Pflanzen, Vol. I–III. Jena: Fischer.
GOLGI, C., 1882–1885: Sulla fina anatomia degli organi centrali del sistema nervoso. Riv.
sper. Frenetria (Reggio-Emilia, Italia) **8**, 165, 361 (1882); **9**, 1, 161, 385 (1883); **11**, 11,
72 (1885).
GOODMAN, H. M., and A. RICH, 1963: Mechanism of polyribosome action during protein
synthesis. Nature **199**, 318.
GOSH, B. K., and R. G. E. MURRAY, 1969: Fractionation and characterization of the plasma
and mesosome membrane of *Listeria monocytogenes*. J. Bacteriol. **97**, 426.
GRASSÉ, P. P., 1957: Ultrastructure, polarité et reproduction de l'appareil de Golgi.
C. R. Acad. Sci. (Paris) **245**, 1278.
GRAY, J., 1955: The movement of sea-urchin spermatozoa. J. Exp. Biol. **32**, 775.
GREEN, P. B., 1962: Mechanism for plant cellular morphogenesis. Science **138**, 1404.
GROVE, S. N., C. E. BRACKER, and D. J. MORRÉ, 1970: An ultrastructural basis for
hyphal tip growth in *Pythium ultimum*. Amer. J. Bot. **57**, 245.
GUNNING, B. E. S., 1965: The greening process in plastids. 1. The structure of the
prolamellar body. Protoplasma **60**, 111.
GUTHRIE, CH., and M. NOMURA, 1968: Initiation of protein synthesis: A critical test of
the 30 S subunit model. Nature **219**, 232.

HANSTEIN, J. v., 1880: Einige Züge aus der Biologie des Protoplasmas. Botan. Abhandl.
aus Morphol. u. Physiol. (Bonn) **4**, Heft 2.
HARRIS, P., 1962: Some structural and functional aspects of the mitotic apparatus in sea
urchin embryos. J. Cell Biol. **14**, 475.
HASSID, W. Z., E. F. NEUFELD, and D. S. FEINGOLD, 1959: Sugar nucleotides in the
interconversion of carbohydrates in higher plants. Proc. Natl. Acad. Sci. U.S. **45**, 905.

HEIDRICH, H. G., 1972: Trägerfreie Elektrophorese zur Isolierung und Charakterisierung von Zellorganellen und biologischen Membranen. (Oral communication January.)

HEPLER, P. K., and W. T. JACKSON, 1968: Microtubules and early stages of cell plate formation in the endosperm of *Haemanthus katherinae*. J. Cell Biol. **38**, 437.

HEYN, A. N. J., 1972: Intra- and extra-cytoplasmic microtubules in coleoptiles of *Avena*. J. Ultrastruct. Res. **40**, 433.

HOLTER, H., 1959: Pinocytosis. Internat. Rev. Cytol. **8**, 481.

HOOKE, R., 1667: Micrographia [or some physiological descriptions of minute bodies made by magnifying glasses]. London.

HUXLEY, H. E., 1957: The double array of filaments in cross striated muscle. J. Biophys. Biochem. Cytol. **3**, 631.

INOUÉ, S., and H. SATO, 1967: Cell motility by labile association of molecules. The nature of mitotic spindle fibers and their role in chromosome movement. J. gen. Physiol. **50**, 259.

— — 1970: Form birefringence of the mitotic spindle. (Oral communication.)

ITO, S., and W. R. LOEWENSTEIN, 1965: Permeability of a nuclear membrane. Science **150**, 909.

JACOB, F., and J. MONOD, 1961: Genetic regulatory mechanisms in the synthesis of proteins. J. Mol. Biol. **3**, 318.

JAMIESON, J. D., and G. E. PALADE, 1967: Intracellular transport of secretory proteins in the pancreatic exocrine cell. J. Cell Biol. **34**, 577, 597.

JAROSCH, R., 1961: Das Characeen-Protoplasma und seine Inhaltskörper. Protoplasma **53**, 34.

JOHN, B., and K. R. LEWIS, 1969: The chromosome cycle. Protoplasmatologia VI B, pp. 1—125. Wien-New York: Springer.

JOHNSON, U. G., and K. R. PORTER, 1968: Fine structure of cell division in *Chlamydomonas Reinhardi*: Basal bodies and microtubules. J. Cell Biol. **38**, 403.

JOST, M., 1965: Die Ultrastruktur von *Oscillatoria rubescens* D. C. Dr. thesis ETH Zürich; Arch. Mikrobiol. **50**, 211.

KAMIYA, N., and K. KURODA, 1965: Contraction of plasmic fibrils. Proc. Jap. Acad. Sci. **41**, 837.

KARLSON, P., 1970: Kurzes Lehrbuch der Biochemie, 7th ed. Stuttgart: G. Thieme.

KIERMAYER, O., 1968: The distribution of microtubules in differentiating cells of *Micrasterias denticulata*. Planta (Berlin) **83**, 223.

KOPP, F., 1971: Ultrastructure of the tonoplast membrane. (Oral communication.)

— 1972: Lokalisation von Membranlipiden im Hefeplasmalemma. Diss. Nr. 4857 ETH-Zürich. Cytobiologie **6**, 287—317.

KORN, E. D., 1969: Cell membranes: structure and synthesis. Ann. Rev. Biochem. **38**, 263.

LEDBETTER, M. C., and K. R. PORTER, 1963: A "microtubule" in plant cell fine structure. J. Cell Biol. **19**, 239.

— — 1964: Morphology of microtubules of plant cells. Science **144**, 872.

— — 1970: Introduction to the fine structure of plant cells. New York: Springer.

LEEDALE, G. F., B. S. C. LEADBEATER, and A. MASSALSKI, 1970: The intracellular origin of flagellar hairs in the Chrysophyceae and Xanthophyceae. J. Cell Sci. **6**, 701.

LENARD, J., and S. J. SINGER, 1968: Reaction of red blood cell membranes with phospholipase C. Science **159**, 738.

LENHOSSÉK, M. v., 1898: Über Flimmerzellen. Verhandl. dtsch. anat. Ges. Kiel **12**, 106.

LEPPER, R., 1956: The plant centrosome and the centrosome blepharoplast homology. Bot. Rev. **22**, 375.

LEWIS, W. H., 1931: Pinocytosis. Bull. Johns Hopkins Hosp. **49**, 17.

LINSKENS, H. F., and J. SCHRAUWEN, 1968: Änderung des Ribosomenmusters während der Meiose. Naturwiss. **55**, 91.

LÓPEZ-SÁEZ, J. F., 1964: Oral communication.

McBride, D. L., and K. Cole, 1969: Ultrastructural characteristics of the vegetative cell of *Smithora naiadum* (Rhodophyta). Phycologia **8**, 177.

Manton, I., 1956: Plant cilia and associated organelles. In: Cellular mechanisms in differentiation and growth (Rudnick, D., ed.), p. 61. Princeton, N. J.: Princeton University Press.

— 1966: Observation on scale production in *Pyramimonas amylifera* Conrad. J. Cell Sci. **1**, 429.

— and K. Harris, 1966: Observation on the microanatomy of the brown flagellate *Sphaleromantis tetragona* Skuja with special reference to the flagellar apparatus and scales. J. Linnean Soc. London **59**, 402.

— D. G. Rayns, and H. Ettl, 1965: Further observations on green flagellates with scaly flagella: the genus *Heteromastix*. J. mar. biol. Ass. U.K. **45**, 241.

Marcus, L., H. Ris, H. O. Halvorson, R. K. Bretthauer, and R. M. Bock, 1967: Occurence, isolation and characterization of polyribosomes in yeast. J. Cell Biol. **34**, 505.

Markham, R., S. Frey, and G. J. Hills, 1963: Methods for the enhancement of image detail and accentuation of structure in electron microscopy. Virology **20**, 88.

Massalski, A., and G. F. Leedale, 1969: Cytology and ultrastructure of the Xanthophyceae. Brit. phycol. J. **4**, 159.

Matile, Ph., 1964: Die Funktion proteolytischer Enzyme bei der Proteinaufnahme durch *Neurospora crassa*. Naturw. **51**, 489.

— 1967: Symposium on yeast protoplasts in Jena 1965. Abh. Dtsch. Akad. Wiss. Berlin, Klasse f. Medizin 1966, Nr. 6. Berlin: Akademie-Verlag.

— 1968 a: Lysosomes in root tip cells in corn seedlings. Planta (Berlin) **79**, 181.

— 1968 b: Aleurone vacuoles as lysosomes. Z. Pflanzenphysiol. **58**, 365.

— 1969 a: Plant lysosomes. In: Lysosomes in biology and pathology (Dingle, J. T., and H. B. Fell, eds.), p. 406. Amsterdam-London: North Holland Publ. Co.

— 1969 b: Reaktionsräume der Pflanzenzelle: Lysosomen und Peroxysomen. Fortschr. Bot. **31**, 64.

— 1969 c: Prospects of yeast cytology. Antonie v. Leeuwenhoek **35**, suppl. 59.

— 1969 d: Enzymologie pflanzlicher Zellkompartimente. Ber. dtsch. bot. Ges. **82**, 397.

— 1971: Spherosomes of maize coleorhiza. (Oral communication.)

— 1972: Hydrolytic activity of isolated lysosomes. (Oral communication.)

— M. Cortat, A. Wiemken, and A. Frey-Wyssling, 1971: Isolation of glucanase-containing particles from budding *Saccharomyces cerevisiae*. Proc. nat. Acad. Sci. **68**, 636.

— M. Jost, and H. Moor, 1965: Intrazelluläre Lokalisation proteolytischer Enzyme von *Neurospora crassa*. II. Identifikation von proteasehaltigen Zellstrukturen. Z. Zellforsch. **68**, 205.

— and H. Moor, 1968: Origin and development of the lysosomal apparatus in root-tip cells. Planta (Berlin) **80**, 159.

— — and K. Mühlethaler, 1967: Isolation and properties of the plasmalemma in yeast. Arch. Mikrobiol. **58**, 201.

— — and C. F. Robinow, 1969: Yeast cytology. In: The Yeasts (Rose, A. H., and J. S. Harrison, eds.), Vol. I, p. 220. London-New York: Academic Press.

— and J. Spichiger, 1968: Lysosomal enzymes in spherosomes (oil droplets) of tobacco endosperm. Z. Pflanzenphysiol. **58**, 277.

— and A. Wiemken, 1967: The vacuole as the lysosome of the yeast cell. Arch. Mikrobiol. **56**, 148.

Maugini, E., and A. Cecchi Fiordi, 1971: Passaggio di materiale dalle cellule del tappeto alla cellula centrale dell'archegonio e al proembrione di *Ginkgo biloba*. Caryologia **23**, 415.

Mayo, M. A., and E. C. Cocking, 1969: Detection of pinocytic activity using selective staining with phosphotungstic acid. Protoplasma **68**, 231.

Mazia, D., 1961: Mitosis and the physiology of cell division. In: The cell (Brachet, J., and A. Mirsky, eds.), Vol. III, p. 77. New York: Academic Press.

— and K. Dan, 1952: The isolation and chemical characterization of the mitotic apparatus of dividing cells. Biol. Bull. (U.S.A.) **103**, 283.

MAZIA, D., P. J. HARRIS, and T. BIBRING, 1960: The multiplicity of the mitotic centers and the time-course of their duplication and separation. J. Biophys. Biochem. Cytol. 7, 1.

MEYER, H. W., and H. WINKELMANN, 1970: Die Darstellung von Lipiden bei der Gefrier-präparation und ihre Beziehung zur Strukturanalyse biologischer Membranen. Exp. Pathol. 4, 47.

MOOR, H., 1964: Die Gefrierfixation lebender Zellen und ihre Anwendung in der Elektro-nenmikroskopie. (Habilitation thesis.) Z. Zellforsch. 62, 546.

— 1967 a: Endoplasmic reticulum as the initiator of bud formation in yeast (S. cerevisiae). Arch. Mikrobiol. 57, 135.

— 1967 b: Der Feinbau der Mikrotubuli in Hefe nach Gefrierätzung. Protoplasma 64, 90.

— and K. MÜHLETHALER, 1963: Fine structure in frozen-etched yeast cells. J. Cell Biol. 17, 609.

— — H. WALDNER, and A. FREY-WYSSLING, 1961: A new freezing-ultramicrotome. J. Biophys. Biochem. Cytol. 10, 1.

MOORE, R. T., and J. H. McALEAR, 1961: Fine structure of mycota. Reconstruction from skipped serial sections of the nuclear envelope and its continuity with the plasma mem-brane. Exp. Cell. Res. 24, 588.

— — 1963: Fine structure of mycota. 4. The occurrence of the Golgi dictyosome in the fungus Neobulgaria pura (Fr.) Petrak. J. Cell Biol. 16, 131.

MORRÉ, D. J., H. H. MOLLENHAUER, and C. E. BRACKER, 1971: Origin and continuity of Golgi apparatus. In: Origin and continuity of cell organelles (REINERT, J., and H. URSPRUNG, eds.), p. 82. Berlin-Heidelberg-New York: Springer.

MOSES, M. J., and J. H. TAYLOR, 1955: Desoxypentose nucleic acid synthesis during micro-sporogenesis in Tradescantia. Exp. Cell Res. 9, 474.

MÜHLETHALER, K., 1971: Studies on freeze-etching of cell membranes. Internat. Rev. Cytol. 31, 1.

— 1972: Electron microscope methods applied to membranes. Chem. Phys. Lipids 8, 259.

— H. MOOR, and J. W. SZARKOWSKI, 1965: The ultrastructure of the chloroplast lamellae. Planta (Berlin) 67, 305.

MÜLLER, W., 1972: Elektronenmikroskopische Untersuchungen zum Formwechsel der Kine-tochoren während der Spermatocytenteilungen von Pales ferruginea (Nematocera). Chromosoma 38, 139.

NAKAO, Y., Y. H. NAKAWISHI, and B. WADA, 1968: Karyokinetic spindles and the behavior of centrioles in the spermatocyte meiosis of silkworms and grasshoppers. Cytologia 33, 125.

NASS, M. M. K., 1971: Stammen die Mitochondrien von Vorläufern der Bakterien ab? Triangel (Sandoz Basel) 10, 29.

NIRENBERG, M. W., 1963: The genetic code II. Sci. American 208, 80.

NORSTOG, K., 1967: Fine structure of the spermatoid of Zamia with special reference to the flagellar apparatus. Amer. J. Bot. 54, 831.

NORTHCOTE, D. H., 1971: The Golgi-apparatus. Endeavour 30, 26.

OUTKA, D. E., and B. C. KLUSS, 1967: The ameba-to-flagellate transformation in Tetramitus rostratus II: Microtubular morphogenesis. J. Cell Biol. 35, 323.

OVERTON, E., 1899: Über die allgemeinen osmotischen Eigenschaften der Zellen, ihre ver-mutlichen Ursachen und ihre Bedeutung für die Physiologie. Vierteljahrsschr. Naturf. Ges. Zürich 44, 88.

PARDEE, A. B., 1968: Membrane transport proteins. Science 162, 632.

PARISH, R. W., 1971 a: Peroxisomes. (Oral communication.)

— 1971 b: The isolation of peroxisomes, mitochondria, and chloroplasts from leaves of spinach beet. Eur. J. Biochem. 22, 423.

PARK, R. B., and N. G. PON, 1961: Correlation of structure with function in Spinacea oleracea chloroplasts. J. Mol. Biol. 3, 1.

PARTHASARATHY, M. V., and K. MÜHLETHALER, 1969: Ultrastructure of protein tubules in differentiating sieve elements. Cytobiol. 1, 17.

Pedersen, H., 1972: The postacrosomal region of the spermatozoa of man and *Macaca arctoides*. J. Ultrastruct. Res. **40**, 366.

Perkins, F. O., 1970: Formation of centriole and centriole-like structures during meiosis and mitosis in *Labyrinthula*. J. Cell Sci. **6**, 629.

Perner, E. S., 1953: Die Sphärosomen (Mikrosomen) pflanzlicher Zellen. Protoplasma **42**, 457.

Pickett-Heaps, J. D., 1966: Incorporation of radioactivity into wheat xylem walls. Planta (Berlin) **71**, 1.

— 1967 a: Further observations on the Golgi apparatus and its functions in cells of the wheat seedling. J. Ultrastruct. Res. **18**, 287.

— 1967 b: The effects of colchicine on the ultrastructure of dividing plant cells, xylem wall differentiation and distribution of cytoplasmic microtubules. Develop. Biol. **15**, 206.

— 1970: Mitosis and autospore formation in the green alga *Kirchneriella lunaris*. Protoplasma **70**, 325.

— and D. H. Northcote, 1966 a: Relationship of cellular organelles to the formation and development of the plant cell wall. J. exp. Bot. **17**, 20.

— — 1966 b: Organization of microtubules and ER during mitosis and cytokinesis in wheat meristems. J. Cell Sci. **1**, 109.

Porter, K. R., 1948: The fine structure of the cytoplasm of cultured tissue cells. Anat. Record **100**, 72.

— A. Claude, and E. F. Fullam, 1945: A study of tissue culture cells by electron microscopy. J. exp. Med. **81**, 233.

— and R. D. Machado, 1960: Form and distribution of ER during mitosis in cells of onion root tip. J. Biophys. Biochem. Cytol. **7**, 167.

— and L. G. Tilney, 1965: Microtubules and intracellular motility. Science **150**, 382.

Pujarniscle, S., 1968: Caractère lysosomal des lutoïdes du latex d'*Hevea brasiliensis* Müll. Arg. Physiol. végét. **6**, 27.

Reisner, A. H., J. Rowe, and H. M. Macindoe, 1968: Structural studies on the ribosomes of *Paramecium*. Evidence for a "primitive" animal ribosome. J. Mol. Biol. **32**, 587.

Renaud, F. L., A. J. Rowe, and I. R. Gibbons, 1968: Some properties of the protein forming the outer fibers of cilia. J. Cell Biol. **36**, 79.

Ringo, D. L., 1967 a: The arrangement of subunits in flagellar fibers. J. Ultrastruct. Res. **17**, 266.

— 1967 b: Flagellar motion and fine structure of the flagellar apparatus in *Chlamydomonas*. J. Cell Biol. **33**, 543.

Robards, A. W., 1968: On the ultrastructure of differentiating secondary xylem in willow. Protoplasma **65**, 449.

— and P. Kidwai, 1969: Vesicular involvement in differentiating vascular cells. New Phytol. **68**, 343.

Robertson, J. D., 1959: The ultrastructure of cell membranes and their derivatives. Biochem. Soc. Symposia No. 16, p. 3. Cambridge, Mass.: Harvard University Press.

— 1960: A molecular theory of cell membrane structure. Verhandl. 4. Internat. Kongreß für Elektronenmikroskopie Berlin 1958, Bd. II, p. 159. Berlin-Göttingen-Heidelberg: Springer.

— 1965: Current problems of unit membrane structure and substructure. In: Intracellular membraneous structure (Seno, S., and E. V. Cowdry, eds.), p. 379. Okayama: Chugoku Press Ltd.

Robinson, D. G., and R. D. Preston, 1972: Plasmalemma structure in relation to microfibril biosynthesis in *Oocystis*. Planta (Berlin) **104**, 234.

Ross, A., 1968: The substructure of centriol subfibers. J. Ultrastruct. Res. **23**, 537.

Roth, L. E., 1967: Electron microscopy of mitosis in *Amebae* III: Cold and urea treatments; a basis for tests of direct effects of mitotic inhibitors on microtubule formation. J. Cell Biol. **34**, 47.

Rouiller, Ch., and E. Fauré-Fremiet, 1958: Structure fine d'un flagellé chrysomonadien: *Chromulina psammobia*. Exp. Cell Res. **14**, 47.

Ruch, F., 1966: Determination of DNA content by microfluorometry. In: Introduction to Quantitative Cytochemistry (Wied, G. L., ed.), pp. 281—294. New York: Academic Press.

Satir, P., 1968: Studies on cilia III: Further studies on the cilium tip and a "sliding filament" model of ciliary motility. J. Cell Biol. **39**, 77.

Schmidt, W. J., 1937: Doppelbrechung von Chromosomen und Kernspindel. Arch. exper. Zellforsch. **19**, 352.

Schnepf, E., 1963: Über den Fangschleim der Insektivoren. Flora **153**, 1.

— 1964: Zur Feinstruktur von *Geosiphon pyriforme*. Ein Versuch zur Deutung cytoplasmatischer Membranen und Kompartimente. Arch. Mikrobiol. **49**, 112.

— 1965: Morphologie der Duftölausscheidung bei *Thyphonium divaricatum* (Araceae). Planta (Berlin) **66**, 374.

— 1969: Membranfluß und Membrantransformation. Ber. dtsch. bot. Ges. **82**, 407.

— and R. M. Brown, 1971: On relationships between endosymbiosis and the origin of plastids and mitochondria. In: Origin and continuity of cell organelles (Reinert, J., and H. Ursprung, eds.), pp. 299—322. Berlin-Heidelberg-New York: Springer.

— and W. Koch, 1966 a: Golgi-Apparat und Wasserausscheidung bei *Glaucocystis*. Z. Pflanzenphysiol. **55**, 97.

— — 1966 b: Über die Entstehung der pulsierenden Vacuolen von *Vacuolaria virescens* (Chloromonadophyceae) aus dem Golgi-Apparat. Arch. Mikrobiol. **54**, 229.

Schrantz, J.-P., 1966: Contribution à l'étude de la formation de la paroi sporale chez *Pustularia cupularis*. C. R. Acad. Sci. (Paris) **262**, 1212.

— 1967: Présence d'un aster au cours des mitoses de l'asque et de la formation des ascospores chez l'ascomycète *Pustularia cupularis*. C. R. Acad. Sci. (Paris) **264**, 1274.

Schwarzenbach, A. M., 1971: Observations on spherosomal membranes. Cytobiol. **4**, 145.

Sievers, A., 1963: Beteiligung des Golgi-Apparates bei der Bildung der Zellwand von Wurzelhaaren. Protoplasma **56**, 188.

— 1967: Elektronenmikroskopische Untersuchungen zur geotropischen Reaktion. II. Die polare Organisation des normal wachsenden Rhizoids von *Chara foetida*. Protoplasma **64**, 225.

Simoni, R. D., M. Levinthal, F. D. Kundig, W. Kundig, B. Anderson, Ph. E. Hartman, and S. Roseman, 1967: Genetic evidence for the role of a bacterial phosphotransferase system in sugar transport. Proc. nat. Acad. Sci. U.S. **58**, 1963.

Sitte, P., 1969 a: Biomembranen: Struktur und Funktion. Ber. dtsch. bot. Ges. **82**, 329.

— 1969 b: Submikroskopische und molekulare Struktur der Zelle. Fortschritte der Botanik **31**, 18.

Sjöstrand, F. S., 1963: A comparison of plasma membrane, cytomembranes, and mitochondrial membrane elements with respect to ultrastructural features. J. Ultrastruct. Res. **9**, 561.

Slautterback, D. B., 1963: Cytoplasmic microtubules. 1. Hydra. J. Cell Biol. **18**, 367.

Sollner, K., 1970: Ion transport across membranes of high ionic selectivity. In: Intestinal absorption of metal ions, trace elements and radionuclides (Skoryna, S. C., and D. Waldron-Edwards, eds.), pp. 21—51. Oxford-New York: Pergamon Press.

Sorokin, H. P., 1967: The spherosomes and the reserve fat in plant cells. Amer. J. Bot. **54**, 1008.

Staehelin, A., 1966: Die Ultrastruktur der Zellwand und des Chloroplasten von *Chlorella*. Dr. thesis ETH Zürich; Z. Zellforsch. **74**, 325.

Stähelin, H., 1954: Über osmotisches Verhalten und Fusion nackter Protoplasten von *Bacterium anthracis*. Schweiz Z. Allg. Pathol. u. Bakteriol. **17**, 296.

Stang-Voss, C., and J. Staubesand, 1971: Mikrotubuläre Formationen in Zisternen des endoplasmatischen Retikulums. (Bindegewebszellen von *Lymnea stagnalis*). Z. Zellforsch. **115**, 69.

Stephens, R. E., 1968: Reassociation of microtubule protein. J. Mol. Biol. **33**, 517.

— 1970: Thermal fractionation of the outer fiber doublet microtubules into A- and B-subfiber components: A- and B-tubulin. J. Mol. Biol. **47**, 353.

Stevens, Ch. L., and M. A. Lauffer, 1965: Polymerization and depolymerization of TMV protein. Biochemistry **4**, 31.

Strugger, S., 1957: Die sublichtmikroskopische Struktur des Cytoplasmas bei verschiedener Fixation. Naturw. **44**, 543.

Thair, B. W., and A. B. Wardrop, 1971: The structure and arrangement of nuclear pores in plant cells. Planta (Berlin) **100**, 1.

Tilney, L. G., 1971: Origin and continuity of microtubules. In: Origin and continuity of cell organelles (Reinert, J., and H. Ursprung, eds.), p. 222. Berlin-Heidelberg-New York: Springer.

— and J. R. Gibbins, 1968: Differential effects of antimitotic agents on the stability and behavior of cytoplasmic and ciliary microtubules. Protoplasma **65**, 167.

— — 1969: Microtubules in the formation and development of the primary mesenchyme in *Arbacia punctulata* II. J. Cell Biol. **41**, 227.

— and K. R. Porter, 1967: Studies on the microtubules in *Heliozoa* II. J. Cell Biol. **34**, 327.

Tolbert, N. E., A. Oeser, R. K. Yamazaki, R. H. Hageman, and T. Kisaki, 1969: A survey of plants for leaf peroxisomes. Plant Physiol. **44**, 135.

Torrey, J. G., 1971: Activities in the Institute of Plant Sciences, Harvard University Cambridge 1969/70. Maria Moors Cabot Foundation for Botanical Research, Report 1969/70, Cambridge, Mass.

Turner, F. R., 1968: An ultrastructural study of plant spermatogenesis: Spermatogenesis in *Nitella*. J. Cell Biol. **37**, 370.

Urban, P., 1969: The fine structure of pronuclear fusion in the coenocytic marine alga *Bryopsis hypnoides*. J. Cell Biol. **42**, 606.

Van der Woude, W. J., D. J. Morré, and C. E. Bracker, 1971: Isolation and characterization of secretory vesicles in germinated pollen of *Lilium longiflorum*. J. Cell Sci. **8**, 331.

Walek-Czernecka, A., 1965: Histochemical demonstration of some hydrolytic enzymes in the spherosomes of plant cells. Acta Soc. Bot. Poloniae **34**, 573.

Watson, J. D., 1963: Involvement of RNA in the synthesis of proteins. Science **140**, 17.

Watson, M. L., 1959: Further observations on the nuclear envelope of the animal cell. J. Biophys. Biochem. Cytol. **6**, 147.

Wehrli, E., K. Mühlethaler, and H. Moor, 1970: Membrane structure as seen with a double replica method for freeze-fracturing. Exp. Cell Res. **59**, 336.

Whaley, W. G., M. Dauwalder, and J. E. Kephart, 1971: Assembly, continuity and exchanges in certain cytoplasmic membrane systems. In: Origin and continuity of cell organelles (Reinert, J., and H. Ursprung, eds.), pp. 1—45. Berlin-Heidelberg-New York: Springer.

— and H. H. Mollenhauer, 1963: The Golgi apparatus and cell plate formation—a postulate. J. Cell Biol. **17**, 216.

Wiemken, A., 1969: Eigenschaften der Hefevakuole. Dr. thesis ETH Zürich.

— 1971: Isolation of yeast vacuoles. (Oral communication.)

Wipf, H.-K., A. Olivier, and W. Simon, 1970: Mechanismus und Selektivität des Alkali-Ionentransportes in Modellmembranen in Gegenwart des Antibiotikums Valinomycin. Helv. Chim. Acta **53**, 1605.

— and W. Simon, 1970: Modelle für Kupplungsmechanismus und Träger-induzierten Alkalitransport in Mitochondrienmembranen. Helv. Chim. Acta **53**, 1732.

Wisniewski, H., M. L. Shelanski, and R. D. Terry, 1968: Effects of mitotic spindle inhibitors on neurotubules and neurofilaments in anterior horn cells. J. Cell. Biol. **38**, 224.

Wohlfarth-Bottermann, K. E., 1963: Grundelemente der Zellstruktur. Naturw. **50**, 237.

— 1967: Microtubules and filaments in *Physarum*. (Oral communication.)

— 1968: Dynamik der Zelle. Mikroskopie **23**, 71.

— and W. Stockem, 1970: Die Regeneration des Plasmalemmas von *Physarum polycephalum*. Wilhelm Roux' Arch. **164**, 321.

Author Index

Subject Index